The Elusive Neutrino

THE ELUSIVE NEUTRINO
A Subatomic Detective Story

Nickolas Solomey

Scientific
American
Library

A division of HPHLP
New York

Jacket and text designer: *Victoria Tomaselli*

Library of Congress Cataloging-in-Publication Data

Solomey, Nickolas.
   The elusive neutrino: a subatomic detective story/Nickolas Solomey.
      p.   cm.—(Scientific American library, ISSN 1040-3213)
   Includes bibliographical references and index.
   ISBN 0-7167-5080-5
   1. Neutrinos—History.   2. Particle nuclear physics.   3. Cosmology.
   I. Title.   II. Series: Scientific
American library series.
QC793.5.N42S658   1997
539.7′215—dc21   96-54844
                                                  CIP

Printed in the United States of America

Scientific American Library
A division of HPHLP
New York

Distributed by W. H. Freeman and Company,
41 Madison Avenue, New York, NY 10010
Houndmills, Basingstoke RG21 6XS, England

First printing 1997, HAW

This book is number 65 of a series.

Contents

To my parents, sisters, and brother

Preface

The Elusive Neutrino first took form as a series of public lectures given in 1994, when I was appointed the 39th Compton Lecturer of Physics at the University of Chicago, Enrico Fermi Institute. I selected the neutrino as my subject, not only because of the particle's allure as one of the most mysterious and elusive elementary particles, but more significantly because the neutrino seemed to be present at so many of the fundamental discoveries that have shaped our understanding of particle physics. By looking at the neutrino, it seemed possible to provide an interesting perspective on the development and substance of particle physics, a relatively new but extremely exciting field of scientific research.

The lectures, presented to a mixed audience of scientists and members of the general public, offered a wonderful chance to refine the presentation of the ideas you will find inside this book. The many excellent questions posed by my listeners suggested ways of improving the organization of the material and its exposition. Although each of the ten chapters still corresponds to a single Compton lecture, some shuffling of chapters and much reformulating of explanations has added to the clarity of the text and the enjoyment to be had from reading it.

The University of Chicago deserves my thanks for giving me the opportunity to present the Compton lectures; these lectures, instituted by the physicist John A. Simpson in 1976, are funded from a bequest by John W. Watzek, Jr. I gratefully acknowledge the encouragement of Vicki Jennings, whose suggestion it was that I transform the lectures into a book, and the unflagging support of Roland Winston, who guided me through the various stages necessary to the preparation of a manuscript that could be submitted for publication. I must also thank my secretary, Judy Gilbert, for her diligent reading and editing of the final draft.

Jenny Brosek and Vicki Jennings, as well as Susan Moran at the Scientific American Library, asked wonderful questions about the physics in each chapter. Their comments as nonscientists greatly assisted my efforts to craft a manuscript that would appeal to the general public. Several physicists graciously shared with me their knowledge of discoveries and events included in these pages; they are Dave Ayres, Samuel Bilinkey, Georges Charpak, Malcolm Derrick, Charles Enz, Ugo Fano, Leon Lederman, P. J. Peebles, Charles Peyrou, Frederick Reines, James P. Rodman, Robert Sachs, Abdus Salam, David Schramm, John Simpson, Jack Steinberger, Earl Swallow, Yau Wah, Bruce Winstein,

Roland Winston, and Tom Ypsilantis. Their willingness to review sections of the manuscript that made use of their remarks helped ensure scientific accuracy. The complete final draft was read by John Bahcall, Amol Dighe, John Ellis, and Robert Wald, and their comments further improved the text's accuracy and coherence.

No book could encompass a complete explanation of all particle physics experiments and theory, certainly not one as brief as *The Elusive Neutrino*. Much that could have been of interest had to be left out, even several topics touching on the story of the neutrino. I did not have space to describe many experiments that confirmed previous results or to mention experiments that attempted to find something new that could not be found or to explain a known result by an unusual process that was not real. Such experiments tend to be forgotten with time, but without them the experimental results that guide us today would not be as clear and certain. My sincere apologies to all those whose work could not be explicitly mentioned in this brief narrative.

Today many hundreds of physicists may contribute to the success of a large experiment. Usually only the spokespeople of experiments achieving important discoveries receive public recognition, but everyone, including the people who build the experimental apparatus, run the accelerator, and operate the computers in charge of data analysis, deserves credit for what they have helped to accomplish. Although I cannot mention all those workers individually, I would like to acknowledge a debt of gratitude, shared by the whole physics community, to them and to the public, governments, and companies who support such work.

In writing *The Elusive Neutrino,* I have tried to share with its readers the physicist's fascination with nature and sense of excitement about solving its mysteries. This fascination seems natural to human beings, whether we are exploring science in the laboratory or in the outdoor world of nature. Modern scientific exploration, with its many fields of inquiry, is really just an extension of the curiosity felt by our ancestors as they stared at the night sky and contemplated our place in this universe and how the world they saw came into being. The neutrino has played a not insignificant role in that continuing endeavor.

Nickolas Solomey
February 1997

The Elusive Neutrino

In the creation of the universe, matter of all sorts was thrown out from an immense explosion called the big bang. This matter clumped into galaxies, some of which are visible in this image produced by the Hubble Space Telescope. Photons and neutrinos produced at this moment of creation are still propagating through the universe; they provide clues to conditions just after the big bang.

CHAPTER 1

The Elusive Neutrino

Tiny particles emanating from the Sun, other stars, and even the big bang that created our universe are propagating through all of space. Thousands are passing through each of us right now without any noticeable effect. These particles, called neutrinos, hardly interact at all with matter in any form; indeed, a neutrino can pass unhindered through several hundred light-years of lead. Judging from our own experiences with this particle, we might expect it to be of little consequence in the workings of the universe. Yet neutrinos produced in the big bang have had a profound influence on the matter that makes up our universe and may even control its eventual fate. Such possibilities were far from the minds of the physicists who first suspected the existence of the neutrino back in the 1930s. These scientists were concerned with understanding the laws of nature and the properties of matter, and they would find that their studies of the neutrino had a central role to play in their investigations.

These modern physicists were continuing an inquiry begun by the ancients many centuries before. The ancient Greeks and Egyptians proposed some surprisingly accurate concepts to describe the physical makeup of the world around them. One great Greek philosopher and mathematician, Democritus, who lived around 430 B.C., believed that all objects were composed of "atoms"—*atomis* being the Greek word for "indivisible part." What he meant was that if the matter in these objects is continuously divided up, at some point one would no longer be able to divide them any further—one would have reached the basic building component. Although this view sounded plausible, the ancient philosophers lacked the technology to conduct experiments that would show their prophecy correct. They believed that nature could be deciphered by simple contemplation without experimentation, a point of view very much the opposite to that of scientists today.

Today we know that all matter is made of atoms, from hydrogen, the lightest, to meitnerium, currently the heaviest atom that has been named. These different types of atoms make up the diversity of objects that we see in the universe around us. The branch of science devoted to the study of the atom, and the properties of matter in general, is physics. Its most fundamental subfield is nuclear and particle physics, the domain explored by those who, like me, seek to understand what makes up the atoms themselves. Near the end of the last century the experimental skills of scientists had developed sufficiently to obtain an answer to the question of what the atom is made of. Since that time we have known that atoms are made of only a few components, which we call subatomic particles. These subatomic components are the electron, proton, and neutron.

The most modern research is not so much concerned with these three particles as with particles that are even more elementary: quarks and leptons, as well as a few more particles that permit these elementary particles to interact. All matter, however different from normal matter, is composed of only a handful of these elementary particles. While the matter we see around us is composed of only two types of quarks and one lepton, stranger forms of matter that use all the quarks and leptons exist elsewhere in the universe—in high-energy cosmic rays, exploding stars, and galactic cores. Here on Earth, accelerators are able to create these unusual forms of matter for scientific study.

The neutrino is one of the most perplexing of these elementary particles. It is a close relative of the electron, but carries no charge and has

extremely little mass if any at all. These are some of the characteristics that make the neutrino so elusive. Neutral particles of substantial mass or charged particles of any mass interact with matter when they encounter it. But the neutrino passes through even our most sophisticated detectors as if it did not recognize that matter was present. The particle's elusiveness contributed to a horrific debate about its existence that broke out when this particle was first proposed. Eventually, as we shall see, physicists managed to prove that this mysterious new elementary particle existed without directly detecting the neutrino itself. Even after its existence had been proved, it took more than thirty years to eventually observe the neutrino in experiments, and even then it was not a neutrino particle itself that was observed, but particles produced from its interaction. Still today the question of whether the neutrino has mass is debated.

Paradoxically, the very property that makes the neutrino difficult to detect also makes it a valuable tool for exploring the secrets of the universe. Because neutrinos rarely interact with matter, they escape unchanged directly from the core of our Sun, where they are produced in the same nuclear reactions that provide Earth with its life-sustaining light energy. The supernova explosion of a dying star produces an even

The Fermilab accelerator, one mile across, seen from above. Immense accelerators of this type smash particles together at high energies to produce new massive particles for experimental study. Neutrinos produced with such accelerators have permitted detailed studies of the electro-weak force.

Four of the scientists who instigated its construction stand in front of the Aleph particle detector at the European Center for Nuclear Research. The large size of the detector is typical of the detectors needed to study particle interactions at high energies. Scientists used this detector to observe electron-positron collisions; from their observations they were able to determine how many types of neutrinos exist in nature.

greater number of neutrinos that can make it to Earth. However, most neutrinos were created billions of years ago in the big bang. Almost all these neutrinos are still propagating through space today, passing through most matter without any detectable effect. By detecting neutrinos, scientists can learn something about what goes on in the interior of our Sun and other stars far away in space, and neutrinos can help us look far back in time to the original creation, permitting scientists to measure parameters of the very early universe moments after the big bang.

The realm of elementary particle physics is so diverse that it can seem overwhelming to casual newcomers who may wish to understand it. The neutrino provides us with an excellent window onto that realm, for that mysterious particle has played a significant role in the history of this field. Using the neutrino as a guide, we will follow the development of modern particle physics from the discovery of radioactivity to current theoretical speculation on the origin of elementary particles and their masses. This is a detective story that begins a hundred years ago, with the discovery of radioactivity. Experimental observations made in those early days presented physicists with certain puzzles that led them to postulate the existence of the neutrino, and along the way they learned much about the laws of physics that guide scientific inquiry.

In particular, as we shall see, investigations of the neutrino have guided physicists in their exploration of one of the four forces through which particles can interact with one another. Understanding these forces is just as essential a part of particle physics as is understanding the composition of the atom.

A force familiar to all of us is the gravity that holds us to the Earth. Although gravity was hypothesized to exist by the ancient Greeks, it has been formulated as a theory expressed in mathematics only within the last three hundred years. The electric force, whose discovery is commonly attributed to Benjamin Franklin's inspiration to fly a kite in a thunderstorm, was known even to the ancient Egyptians, although they used it mostly for magic tricks, as a curiosity, or perhaps to ward off thieves. Nevertheless, a scientific formulation of this force has been achieved only within the last 150 years. Still more recently, at the end of World War II, our planet saw for the first time the release of a tremendous violent energy that only the subatomic, so-called strong nuclear force could make possible. The explosion of the atomic bomb was a clear demonstration that our understanding of the composition of matter in the universe, and the forces through which it interacts, had reached a new level. Experiments with the neutrino have been essential to revealing a new force, the weak nuclear, that is now known to exist. This is the force causing one of the types of radioactivity that we shall encounter. Although the smaller of the two nuclear forces, it would help to explain much about what we know regarding the constituents of matter within the proton and neutron.

As scientists learned more about particle physics, they found the neutrino to be more perplexing than originally thought. We shall see that the particle was found to violate a precious conservation law that had been believed to be a cornerstone in physics, and in consequence that law had to be discarded. Furthermore, the neutrino may be the only fundamental particle of nature that is its own antiparticle, giving it a special property that other forms of matter cannot possibly possess. These, and the many other properties of the neutrino and of the forces through which it interacts, are fascinating to physicists. This fascination keeps us searching to understand what makes up the world we live in. In this book I hope not only to show the theoretical development that is so much a part of elementary particle physics, but to dive deeply into stories of experiments and their results, which truly show how we know that these particles and their many mysteries are real. These two seemingly different aspects of science, theory and experiment, are inseparable from each other in modern science, and further advancement in scientific understanding strongly depends on both.

The Discovery of Radioactivity

More than a hundred years ago, common household objects were sometimes made of special colored glass that had surprisingly beautiful colors of blue and orange. These eerie colored housewares were manufactured by adding to glass small quantities of special ores containing heavy metals such as cobalt and uranium. Although present-day consumers would be leery of buying drinking glasses, vases, or coffee tables with the descriptive names cobalt-blue or uranium-orange, the decorative style was a common one of the period. But soon these elements, and more yet to be discovered, would take on a special meaning in the scientific community.

In 1896, the French physicist Henri Becquerel accidentally noticed that ore samples containing these heavy metals had a peculiar effect. When he stored unexposed photographic plates near the samples, the plates, acting as if they had been directly exposed to light, turned black. These ores seemed to be emitting some kind of new energy, an energy with at least some properties in common with light. Becquerel's observation marks the birth of modern nuclear physics, for he was the first to notice the phenomenon that we now call the radioactive decay of atoms.

Becquerel knew that his ore samples contained the element uranium. A few years after his accidental discovery, Marie and Pierre Curie showed after careful study that these peculiar ores also contained

A glass sugar bowl doped with a small amount of uranium gives an eerie orange-yellow color. This sugar bowl dates from about 1850.

Pierre and Marie Curie, shown in their laboratory shortly after their discovery of the element radium in 1898.

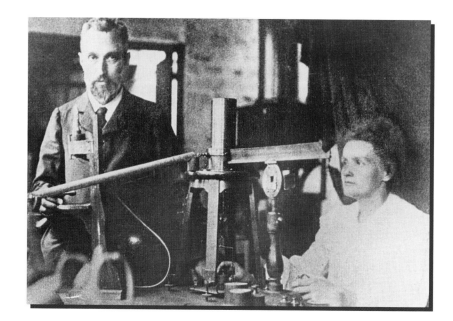

thorium and the previously unknown element radium. The Curies and other scientists demonstrated that all three of these materials underwent a kind of transition, which the Curies called radioactive decay. This process not only transforms an atom into an entirely different kind of atom of a different weight, but also releases particles of tremendous energy.

A seemingly simple experiment revealed that radioactive elements can emit three different types of particles, called alpha, beta, and gamma. A sample of radioactive ore stored in a vessel pierced only by a narrow slit, as depicted in the figure on page 7, emits particles in a narrow beam. When the beam is passed through a magnetic field, the positive alpha particles in the beam bend one way and the negative beta particles bend in the opposite direction. Neutral gamma particles pass through undeflected. Today we know that an alpha particle is a helium nucleus, of positive charge, being ejected from a heavier nucleus. A beta particle is an electron with negative charge, and gamma rays are high-energy photons—discrete packets of light that exhibit characteristics of both a particle and wave. Most samples of radioactive ore are mixtures of many different elements, which is why they appear to emit all three types of particles; a pure element emits mostly one type. Scientists, recognizing that the high-energy particles emitted by radioactive elements presented them with a powerful research tool, began to design experiments that explored the fundamental constituents of the atom itself.

In England, just before the turn of the century and about the time that radioactive elements were discovered, J. J. Thompson discovered the electron, prompting him to propose a model of the atom called the plum-pudding model. He envisioned a sea of positive charge with electrons of negative charge distributed throughout, like the plums in a pudding. At that time such a model seemed reasonable because it explained how the electron of negative charge combined with a positive charge to form the neutral, uncharged atom. The plum-pudding model of the atom was only speculation, but it was an idea that could be readily tested.

In 1911, Thompson's student, Ernest Rutherford, proceeded to prove his mentor's model of the atom wrong. Rutherford designed an experiment that he hoped would reveal the distribution of charge in an atom. He aimed a collimated beam of radioactively emitted alpha particles at a thin foil of gold atoms, behind which he had placed a fluorescent screen that would allow the scattering of alpha particles to be observed. According to the electrostatic laws of physics, similar charges

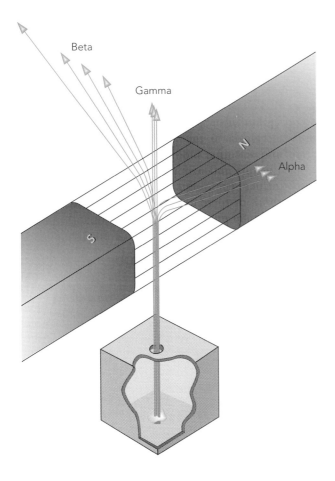

The alpha particles and beta rays emitted by a sample of radioactive ore are deflected in opposite directions in a magnetic field, while gamma rays continue undeflected on a straight path.

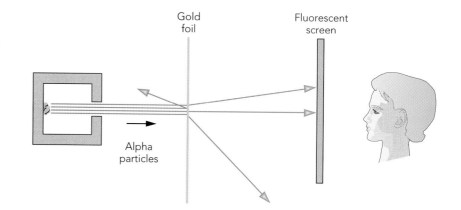

Rutherford's alpha-particle scattering experiment. Energetic particles from the radioactive source impinge on the gold foil, and when these particles hit the fluorescent screen their scattering angle can be observed. The source could be rotated in relation to the screen to provide observations of more dramatic scattering at even larger angles.

repel each other, so as the positive alpha particles passed through the foil and came close to the positive charge in an atom of gold, they should be deflected. If the positive charge was distributed evenly in the atom, as Thompson's model predicted, then the paths of the alpha particles would be only slightly bent from their initial trajectories. However, some alpha particles were deflected by more than a right angle, and, more dramatically, other particles bounced back entirely. Rutherford joked that it was like firing a cannonball at a wall and having it come right back at him. The experiment was run many times, always with the same result. There could be little doubt as to the meaning of these results: the positive charge of an atom must be highly concentrated.

In the early days of this century, experimenting with particles was difficult; the only way to observe alpha-particle scattering on a fluorescent screen was with the naked eye. To become a student of Rutherford's one had to be more than just a bright pupil. The candidate student would be required to sit for several hours in a dark box, one side of which was a fluorescent screen. Then, after the student's eyes had adapted to the low light level, a known intensity of light from alpha particles was impinged on the fluorescent screen; the candidate student had to be able to see reliably a flash of 6 photons of visible light made by the impinging alpha particle (the average person is capable of seeing only flashes of light with at least 10 photons). All of Rutherford's famous experiments with beams of alpha particles relied on sharp-eyed observers to count the flashes.

One of Rutherford's more famous students was Hans Geiger, who, to make it easier for himself to perform his teacher's scattering experiments, invented an instrument that electronically recorded the passage of particles. This useful electronic counter, referred to generically as a Geiger counter, is still widely used today to measure levels of radiation.

Hans Geiger and Ernest Rutherford in their laboratory at Manchester University, England, in about 1908. With them is the apparatus they used in their alpha-particle scattering experiments.

The distribution of alpha particles on the fluorescent screen revealed for the first time that the positive charge distribution in an atom is concentrated, but a similar concentration of negative charge was not observed. Rutherford postulated that the atom is similar in construction to the solar system. The atom, he suggested, is composed of a massive core called the nucleus, which is surrounded by negative electrons that orbit in planetary-like paths. The electrons are bound to the positive core through the electromagnetic force, a force that causes opposite charges to attract.

Breaking the Law

The science of physics is based on a few presumed laws that appear to be obeyed no matter what system is being investigated. One of these is the law of conservation of energy, which tells us that the total energy of a system, both before and after any given instant in time, is always the same. A ball and a hilly road provide just such a system. If a ball is let go from the top of a hilly road, it will roll down to the bottom of the hill and then start going up the other side. The law of conservation of energy tells us that the ball cannot go any higher than the height from which it was released. To go higher, energy must be applied to the ball either at its release, by giving it a hard push, or through some other

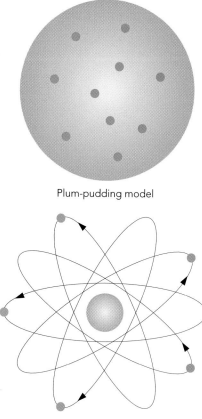

Plum-pudding model

Rutherford model

Models of the atom based on experimental results. J. J. Thompson's plum-pudding model has electrons in a sea of positive charge, whereas Rutherford's model shows electrons orbiting a heavy core of concentrated positive charge.

means at any point along its path. Similarly, if the ball does not reach a height equal to the height from which it was released, then energy must have been lost along the path, a loss most likely due to friction. The instant after the ball is released, it has no speed and all of its energy is *potential*. Gravity acts upon the unrestrained ball and causes it to start rolling down the hill. At the bottom of the hill the ball is rolling quickly: all of the potential energy has been transformed into *kinetic energy*, also called energy of motion. The kinetic energy of the ball at the bottom of the hill is equal to the potential energy that was lost when the ball went from the top to the bottom; thus, energy is conserved.

Sometimes energy appears in other forms. When a book is held and then dropped onto a tabletop, the instant before impact it will have acquired an energy of motion dependent on whatever distance it fell. After it hits the table, however, it is obviously not moving and has no kinetic energy of motion. It also has no potential energy since it has already fallen. The falling book may at first appear to violate the conservation of energy law, but in reality it does not. When the book slammed onto the table it created a noise, which took energy; in addition, both the book and the table became slightly heated because some of the energy was converted from motion into heat.

Another fundamental law is the conservation of momentum; here momentum is the product of the mass of an object and its velocity. Like energy, the momentum in a system of particles is always conserved; the sum of momentum before and after any collision is always the same. Keep in mind that just as with energy conservation, external forces can superficially make this law appear to be violated. A nice example of the conservation of momentum comes from American-style football—specifically, the head-on collision of two front linemen. If both have the same mass and are traveling at the same speed when they collide head on, then they will both come to a stop at the collision point. Their momentum, being equal but opposite, sums to zero after the collision. However, if one lineman is traveling faster than the other, then after the collision both entangled linemen will move in the direction of the faster, although at a much reduced speed. The mass of the linemen plays an equally important role. A more massive object is harder to stop, so when two linemen collide head on at the same speed, the heavier one will control the direction of motion after the collision, but again he will not move as fast as he did before the collision.

When scientists began to study radioactive beta decays in more detail, they were shocked to find that these decays appeared to break these laws of conservation. They were especially puzzled by the apparent violation of the conservation of energy law. When matter decays, its initial mass does not equal the mass of the decay particles. Since mass is a form of potential energy, this was a discrepancy that needed some explanation.

Einstein's famous mass-energy equation,

$$E = mc^2,$$

describes the equivalence of mass with energy. The equation states that the mass m is equivalent to a form of potential energy E, while c, the speed of light, serves as a conversion factor. Particle physicists often describe mass in terms of energy. The basic unit is the electron volt, abbreviated eV, which is the energy an electron has when passed through a potential difference of one volt. For instance, when physicists say that the proton's mass is 938 MeV/c^2, what we actually mean is that the mass of the proton, if it is converted into pure energy, is 938 MeV, where the symbol M stands for million. For comparison's sake, the electron's mass is only 0.511 MeV/c^2, nearly 2000 times lighter than the proton. Always implicit in any discussion about the mass of a particle is that the kind of mass being discussed is the rest mass, the mass of the particle when it is not moving. Other types of masses, such as the "relativistic" mass of a particle moving near the speed of light, become important in special physical circumstances.

In equating mass with energy, Einstein's famous equation provides a possible explanation for the seemingly lawless behavior of beta decay: the mass lost from the initial particles has been converted to energy.

What happens to that mass is best illustrated by alpha decay, the emission of an alpha particle by a nucleus. Although some of the energy of the radioactive decay goes into the recoiling of the remaining nucleus, most of the energy is imparted to the lighter particle—in this case, the emitted, positive alpha particle. The energy of the alpha particle accounts for all the mass that seems to disappear during decay. Conservation of energy, therefore, remains the law of the land.

This explanation, which worked so neatly for alpha particles, did not solve the difficulties raised by beta decay. The beta-decay emission of an electron has an unusual energy distribution, one that is quite different from the energy distribution observed with alpha decays. This energy distribution caused concern among physicists, who encountered many problems in trying to explain it.

In 1913, just prior to the outbreak of World War I, James Chadwick—an Englishman studying in Berlin, where Geiger had become director of the Physics Institute—was trying to measure the energy of electrons emitted in beta decay. Chadwick passed electrons from beta decay through a magnetic field. The degree to which an electron was deflected by the field would reveal, with the help of a few calculations, the energy of the electron. Chadwick's measurements clearly showed that the electron's energy was not the same well-defined value in every case, as in alpha decay; instead he recorded a continuous, broad distribution of values. That is, different beta decays of the same types of nuclei were

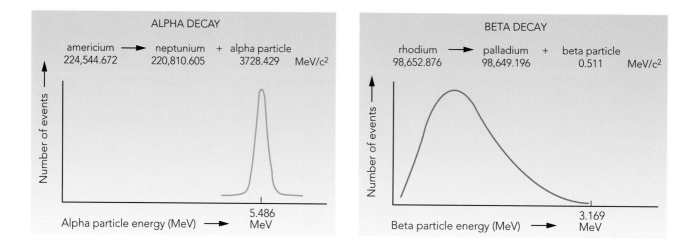

ALPHA DECAY

americium ⟶ neptunium + alpha particle
224,544.672   220,810.605   3728.429   MeV/c²

Number of events →

Alpha particle energy (MeV) ⟶   5.486
                                   MeV

BETA DECAY

rhodium ⟶ palladium + beta particle
98,652.876   98,649.196   0.511   MeV/c²

Number of events →

Beta particle energy (MeV) ⟶   3.169
                                  MeV

The measured energy of particles emitted in the radioactive decay of the element americium by alpha decay and the element rhodium by beta decay. The alpha particle carries away a well-defined amount of energy in each decay, while the beta decay has a broad energy distribution with much missing energy that initially could not be accounted for.

producing beta particles with different energies! Although Chadwick performed this experiment many times with a variety of radioactive beta-emitters and although he measured the electron's energy by two separate means—by a method that counted electrons after deflection and a method that measured ionization energy—the result was always the same: a puzzling broad spectrum. Why the beta decay's electron-energy distribution was not like alpha decay's—a single well-defined value—was not decipherable based on the known laws of physics. Especially worrisome was the apparent violation of the conservation of energy principle: in every case the decaying particles started out with the same potential energy; if energy was conserved, how could the emitted particles end up with differing observed energies? This conundrum would avoid a suitable explanation for fifteen years.

An explanation that was suggested soon after Chadwick produced his curious spectrum was that electrons could be losing energy in differing amounts as they passed through the material containing the radioactive source. A radioactive source is never pure; a sample of radioactive material contains not only the material of other radioactive nuclei that have not yet decayed, but also stable nuclei as well as nuclei remaining from previous decays. The emitted beta particles could interact with these other nuclei, causing the beta particles to lose energy in differing amounts depending upon how many interactions occurred before they exited the source. This explanation was plausible, given that electrons were known to interact more readily with matter than could positive alpha particles, and it would explain the distortion of the electron spectrum.

In an attempt to prove this hypothesis, in 1927 Charles D. Ellis and William A. Wooster set up an experiment that measured the total energy of a decay. Chadwick's experiment had measured only the ki-netic energy of an electron and missed any energy dissipated as heat. To obtain a measurement for energy lost as heat, Ellis and Wooster placed their radioactive sample in a heat calorimeter, which is a well-insulated vessel in which the amount of heat could be measured with sensitive electronic thermometers. Any electron energy that the magnetic analysis of the decay did not observe would be measured as heat. Ellis and Wooster first tested their equipment's sensitivity by measuring the heat emitted from an electrical wire that had a well-known current passing through it; the system proved to be trustworthy.

When an intense beta-emitter was lowered into the calorimeter for a fixed amount of time, the instrument measured the total energy carried by any interacting particle coming from radioactive decays. If the beta-decay electrons were emitted with a fixed amount of energy, as in alpha decay, but lost some of that energy in differing amounts because of the way in which individual electrons exited the source, the values mea-sured should be equivalent to the highest energy measurements of Chadwick's electron-energy spectrum. Instead, the value measured was closer to the mean. This result effectively eliminated the possibility that the electron-energy distribution from beta decay was a distortion caused by the electron's passage through matter. A proper explanation of the cause of this phenomenon remained elusive.

The problem presented by beta decay was not limited merely to the discrepancy of the electron-energy distribution. Rutherford's alpha-scat-tering experiments had clearly demonstrated that the nucleus carries a positive charge, that the negatively charged electrons orbit the nucleus, and that an element can be identified by the number of positive charges, or protons, in its nucleus. During alpha decay, a nucleus loses two posi-tive charges; during the neutral gamma decay, a nucleus does not change; but in beta decay, the nucleus *increases* by one positive charge in association with the emission of an electron. It appears, then, that the electron is being emitted from the nucleus itself!

Solving the Problem

Faced with this problem, Wolfgang Pauli—an outstanding professor of theoretical physics at the Federal Technical Institute (ETH) in Zürich, Switzerland—proposed that there could be only two ways to explain the electron-energy distribution of beta decay. Either the principle of en-ergy conservation is only valid statistically over many events and not at the level of an individual atomic decay, or energy conservation is exact and another particle is emitted along with the electron, a particle that cannot be observed but that would carry away some of the energy of the decay. This proposal put Pauli in an uncomfortable position. After other physicists had circulated his speculation about an unobservable

Wolfgang Pauli of ETH Zürich lecturing on his theories of quantum mechanics in Copenhagen in 1929. Pauli suggested that an elusive, unobserved particle was present in beta decay.

particle, Pauli was extremely embarrassed that his name was attached to the proposal. He felt that if his proposal was wrong, he would be defacing the value of his great achievements in quantum mechanics, for which he would eventually win the Nobel Prize in 1945. He has been quoted as saying to a friend, "I have done something very bad today by proposing a particle that cannot be detected; it is something no theorist should ever do." The particle, of course, is the subject of this book—the elusive neutrino.

Finally, after much agonizing about whether he should openly suggest the neutrino, on the fourth of December, 1930, Pauli wrote a letter of introduction to a conference organizer imploring him to listen to what the courier of his letter had to say about the proposition for the neutrino's existence, and how it could help explain the mystery of the electron emitted in beta decay. The letter proposed that there were more particles in the nucleus of an atom than just the positively charged protons. These particles were called neutrons (their actual discovery would soon be announced by James Chadwick, working in Rutherford's lab). It was the neutron that decayed into an electron and a neutrino, both of which left the nucleus, and a positively charged proton that remained in the nucleus. Pauli expressed this idea in a charming letter, reproduced on the opposite page, although note that in his letter he did not distinguish between the neutron and neutrino, which we now know to be completely different particles. Pauli did not attend the conference but, instead, excused himself in order to attend a village ball. (After all, what is more important, a village ball or a conference announcing the possible existence of a new particle?) Conveniently, Pauli had found a way to present the neutrino to the world without tarnishing the reputation that he prized so highly. Still, Pauli was uncertain whether the neutrino could ever, really, be proven to exist.

Even though Pauli was the first to conceive of the neutrino, it was a young Italian physicist by the name of Enrico Fermi who first brought the particle into the realm of reality. Fermi, who already had a Doctor of Science degree from the University of Pisa, had been spending his summers away from Italy in order to study. During the summers of the late 1920s he learned the new quantum mechanics from Max Born while in Göttingen, Germany. During the summers of the 1930s, while in Brussels, Belgium, he learned the new theoretical techniques of quantum electrodynamics, which would later evolve to become known as QED. Quantum mechanics and QED are modern theories of how charged particles, such as electrons, interact with an electric field. These studies abroad prepared Fermi for the challenges posed by the problem of beta decay.

Within the framework of quantum electrodynamics, Fermi worked toward an explanation of the puzzling electron-energy distribution of beta decay. He thought that a neutron could decay into an electron and the recently discovered antielectron, called the positron, *if* the positron could be captured by the nucleus somehow. Because Pauli remained skeptical about the neutrino's existence, he never published a paper about it, but that didn't stop him from guiding other physicists in their thinking about the problem with beta decay. Pauli convinced Fermi that this suggestion was not the proper one. Pauli suspected that if the neutrino was involved and a proper theory of its involvement could be formulated, it might be possible to confirm the particle's existence without detecting it directly.

It was Enrico Fermi, then, who ultimately developed a "four-point beta-decay" theory that mathematically described the beta-decay energy spectrum. Using the new quantum electrodynamics and incorporating Pauli's suggested neutrino as the culprit that carried away the missing energy, undetected, Fermi's mathematical formulation of the theory showed that it is the neutron—also bound to the proton in the atomic nucleus—that decays into a proton, and that simultaneously emits the electron and neutrino. Fermi's theory is called the four-point theory because, since one particle exists before the decay and three afterward, four particles come together at the point of transformation. The existence of a third particle emitted in the beta-decay process provided an explanation for the broad energy distribution of the electron: the unobserved particle could take away a different share of the energy with each decay, sometimes leaving less and sometimes more for the electron. It

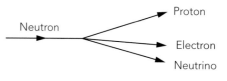

Fermi's four-point beta-decay theory proposed that during beta decay the neutron decays into a proton, an electron, and a neutrino. The escaping neutrino provides an explanation for the missing energy seen in the beta-decay spectrum.

Physikalisches Institut
der Eidg. Technischen Hochschule
Zürich

Zürich 4 dec. 1930
Gloriastr.

Dear Radioactive Ladies and Gentlemen

As the bearer of these lines will explain to you in more detail—and I beg you to listen to him with benevolence—I have considered, in connection with the 'wrong' statistics of ^{14}N and ^{6}Li as well as with the continuous β spectrum, a way out for saving the 'law of change' of statistics and the conservation of energy: i.e. the possibility that inside the nuclei there are particles electrically neutral, that I will call *neutrons,* which have spin 1/2 and follow the exclusion principle and that in addition differ from photons because they do not move with the velocity of light. The mass of the neutrons should be of the same order of magnitude of that of the electrons and anyhow not greater than 0.01 protonic masses. The continuous β spectrum would then be understandable, assuming that in the β decay together with the electron, in all cases, also a neutron is emitted, in such a way that the sum of the energy of the neutron and of the electron remains constant.

The question is now to see which forces act on the neutrons. The most probable model appears to me to be, for wave mechanical reasons (the detail can be given to you by the bearer of these lines), for the neutron at rest to be a magnetic dipole of a certain moment μ. The experimental data certainly require for the ionizing power of such a neutron to be not greater than that of a gamma ray and therefore μ should not be greater than $e \times 10^{-13}$ cm. I do not consider advisable, for the moment, to publish something about these ideas and first I apply to you with confidence, dear Radioactives, with the question: what do you think about the possibility of providing the experimental proof of such a neutron, if it would possess a penetrating power equal or ten times greater of that of gamma rays?

I admit that my solution may appear to you not very probable, because if the neutrons would exist, they would have been observed long since. But only who dares wins, and the gravity of the situation in regard to the continuous β spectrum is enlightened by the opinion of my predecessor in the chair Mr. Debye, who long since told me in Brussels: 'Oh, the best thing to do is not to talk about, like for new taxes'. For this reason one should consider seriously any way towards safety. Thus, dear Radioactives, consider and judge. Unfortunately I cannot come personally to Tübingen, because I am necessary here for a ball that will take place in Zürich the night from 6 to 7 December.

With many greetings to you as well as to Mr. Back.
Your devoted servant,

W. Pauli

was Fermi who named the unobserved particle in beta decay "neutrino." In Italian, neutrino means "little neutral one."

Back in the early days of particle physics, this particle was referred to as a neutrino; however, it is technically an antineutrino. It took physicists a long time to understand this, and in this book I will not have introduced the concepts necessary to understand this difference until the end of Chapter 5. It is a convention in physics to talk about antineutrinos using the simpler word neutrino, just as it is common practice to say that a particle's mass is 1 GeV, when we really mean that its mass is 1 GeV/c^2, where G stands for billion. Until the concepts of neutrino and antineutrino are straightened out, I will refer to both types of particles as neutrinos, and in the figures of these first four chapters the symbol for the neutrino will not be used, but the word "neutrino" will be written out.

After Fermi finished writing the paper describing his theory, he invited some friends to his room. The excited young Fermi, sitting in the middle of his bed, read his paper to several close friends and colleagues—Emilio Segré, Edoardo Amaldi, and Franco Rasetti—who were sitting around him on the edge of the bed. As he read, Fermi joked offhandedly about how this paper was by far his most important, and contained the work that would make his mark in history. Ironically, this

The young theoretical physicist Enrico Fermi (left) taking a rest while mountain climbing in the Alps with friends Nello Carrara (middle) and Franco Rasetti (right) in 1922.

was the first and last theoretical work that Fermi would do on beta decay, and it is the theoretical work for which he is best known. The actual work for which he is probably most famous, and its impact on the direct detection of the neutrino, will be discussed in Chapter 4.

The success of Fermi's four-point beta-decay theory should not be understated. It conclusively resolved a significant controversy about the nature of the atom and explained why and how an electron is emitted from a nucleus. Most important, Fermi's theoretical model matched the experimental observations. Even though the neutrino had not yet been directly detected, its physical effects on the observed electron-energy distribution could be seen, and they agreed with the newly developed theory. Without an agreement between theory and experiment, neither can really be trusted. A theory that cannot be confirmed by experiment is reminiscent of the beliefs of the ancient Greek philosophers, who had no need for experimental investigation and believed that all of nature could be discovered through philosophical inquiry—a belief that is antithetical to science.

In 1934, Fermi's paper on the theory of beta decay was published, and it is generally regarded as a turning point that would eventually usher in a new field of study within the broader discipline of nuclear and particle physics. This new field was the study of the weak nuclear force; this force was first noticed in beta decay, and many advances in our understanding of it emerged from the study of the neutrino. Today we know that there are four forces by which matter interacts: *gravity,* the interaction of mass; *electromagnetism,* the interaction of charge; the *strong nuclear force,* the interaction that binds protons and neutrons in nuclei; and, with the understanding of the rudiments of beta decay, we now have the *weak nuclear force.*

In the 1930s, little work on subatomic particle physics was being done in the United States. Albert Einstein was at Princeton University, but he was so convinced that energy conservation was absolute that he would not consider investigating any problem that suggested it might be otherwise. Although, ultimately, beta decay was shown to obey the fundamental laws of physics, had not alternatives to the fundamental laws been investigated some of nature's deeply buried secrets might have remained unknown for a long time.

. . .

In this chapter we have seen how a relatively small problem with the electron-energy distribution of beta decay could not be understood in terms of known principles of physics. To solve this problem, Wolfgang Pauli proposed that a particle, the as yet undetected neutrino, might carry away some of the energy. Without the proposed neutrino to carry away energy, the law of conservation of energy would have been called into question, a highly unacceptable development. Enrico Fermi developed the first theory that used the proposed neutrino to correctly describe the unusual electron-energy distribution of beta decay.

From a seemingly small problem, a whole new field of particle physics emerged. Whereas the theory of quantum mechanics was formulated over a short period of time, our understanding of interactions mediated by the weak nuclear force has been developing over the past sixty years and is still not complete.

An image produced by an infrared telescope reveals the high density of stars at the center of our galaxy. Since these stars formed and ignited, they have produced a large number of neutrinos, which continue to propagate outward.

CHAPTER 2

Cosmology and the Neutrino Mass

One of the first things that Fermi's new theory of beta decay provided was a way to measure the neutrino's mass. Indeed, it was in theory possible to measure that mass before the neutrino itself had been detected directly. The neutrino's ability to carry away a large amount of energy completely undetected implied that its mass was substantially smaller than the electron's. The neutrino could even be a massless particle. Yet in contrast to what we know about the photon, which is required to have a zero mass by the fundamental theories of quantum electrodynamics, there is no known theory that requires the neutrino to be massless. For many years the neutrino appeared to have a small but nonzero mass, and the uncertainty over its mass has added to the lure of its mystery.

Lately scientists have had a special reason to be interested in the neutrino's mass. If that mass were determined to be nonzero within particular parameters, an intimate connection could be made between the neutrino and the ultimate fate of our universe. Attempts to predict the eventual fate of the universe belong to the realm of science called cosmology. These two seemingly unrelated topics, the neutrino's mass and cosmology, have now crossed paths, but before we can see how, I shall take a moment for a brief introduction to cosmology, encompassing only some essential details of the subject.

The Historical Origins of Cosmology

When gazing at the night sky, most of us first notice the many stars and constellations. Although human beings have probably always wondered at the simple beauty of these evening gems, we have lately taken that wonder a step further. The origin of stars, especially what it might say about our place in this universe, has been the subject of philosophical contemplation and scientific inquiry by most civilizations throughout history. Now specialists in the field of cosmology are studying the growth and development of the entire universe, which contains all matter in existence, both visible and invisible. Their goal is to discover, through scientific means, the origin and eventual fate of the universe.

Beginning in 1609, Galileo's improvements to the telescope—originally invented so that merchants on shore could identify approaching commercial vessels—allowed astronomers to observe rich detail in what had previously appeared to be simple points of light in the night sky; they were also able to identify hundreds of new stars that the naked eye could not detect. One of Galileo's most profound discoveries came as a result of his careful observations of the moons of Jupiter and the phases of Venus; these studies provided him with proof that the Sun, and not the Earth, is at the center of our solar system. Because of their religious beliefs, the people of that time were not prepared to accept that the Earth isn't the center of our universe, and Galileo was denounced by the Inquisition and banished for the last nine years of his life.

Through observation, astronomers learned that stars clump together to form galaxies; these come in many varieties, the most spectacular of which are the spiral galaxies, with their long, sweeping arms. On a moonless night, an observer standing atop a remote mountain far from city lights can see the arms of our own galaxy, the Milky Way, as a straight band of diffuse light across the sky.

An elegant spiral galaxy, NGC 2997, viewed from the top. Its dense core and giant spiraling arms are clearly visible.

Stars are held together in galaxies by the mutual gravitational attraction that they exert upon one another. The galaxies themselves are most often found clumped together in what are called galactic clusters (a simple but accurate term). When astronomers began to map the galaxies in the farthest reaches of space accessible to their telescopes, they noticed that, in every direction they looked, the galaxies and even the clusters of galaxies seemed to be uniformly distributed. This simple observation was extrapolated to be true everywhere in the universe. This formulation, called the cosmological principle, states that on a large enough scale, the universe is both homogeneous and isotropic; that is, the distribution of matter is the same everywhere and the universe looks everywhere the same. The cosmological principle has been used to imply that, because the universe is the same everywhere, it has no center.

Studying the development of a star or galaxy is difficult. These ancient and slowly changing objects do not grow in any appreciable way during the average human lifetime. Luckily, light takes time to travel to us; when we observe the galaxies closest to us we are observing those that are approximately our own age, but the more distant the galaxy, the further back in time we look. This is true provided all the galaxies and stars we are looking at formed at about the same time. This conjecture seems to be a reasonable one and allows us to look back in time to a limited degree. Assuming that the cosmological principle is correct, we can also assert that objects in close proximity to us, which are easier to study, are representative of the rest of the universe. This assertion allows us to formulate theories about the universe, even those parts that are impossible to observe.

Doing experiments in cosmology is an almost impossible task; hence cosmology is mostly an observational science. A description of one observational technique, measuring Doppler shift, will give us an appreciation for astrophysicists' cleverness with simple measurements. The Doppler shift has revealed a fundamental mystery about the universe, one that has yet to be solved, but whose solution was once thought to involve the neutrino.

The physical phenomenon called Doppler shift is something that most of us have observed, even if we might not have known there was a name for it. Doppler shift allows an observer to determine whether a source of waves is moving toward or away from her. Sound is a kind of wave, for example. If someone hears a jet plane overhead, the pitch of the engine rises as the plane comes toward the listener and then falls as the plane moves away. The motion of the plane causes the wavelength of the sound wave to seem shorter as it approaches the listener, producing a higher-pitched sound, and longer as it moves away from the listener, producing a lower-pitched sound. The jet engine frequencies heard by the listener are different from the true frequency produced by

From the perspective of the listener, the wavelength of sound waves shifts according to whether listener and emitter are moving farther apart (middle) or closer together (bottom). Only when listener and emitter are not in motion with respect to each other do sound waves remain unshifted (top). For sound waves the difference in wavelength in each case causes a noticeable difference in pitch. The emitted light from the wingtips of an airplane moving at speeds close to that of light would cause an observable red shift if the plane were moving away from the observer and an observable blue shift if the plane were moving toward the observer.

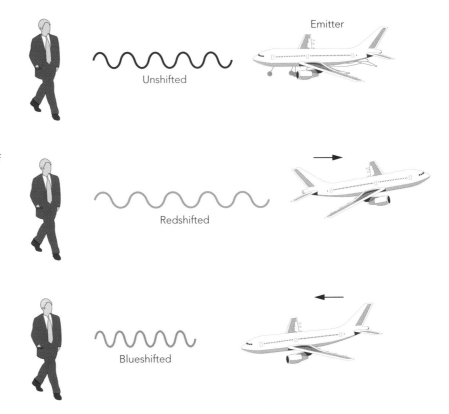

Emitter

Unshifted

Redshifted

Blueshifted

the engine, which would have been heard only when the plane was directly overhead.

To astronomers and cosmologists, it is light, not sound, that is important. If a light-emitting object and its observer are always at a fixed distance from each other, the wavelength of the light will appear to remain the same at the location of the observer as it is at the location of the emitting object, just as the sound of the jet would have stayed the same if the jet had not been in motion. When the object begins to move away from the observer, the wavelength of light appears to stretch out. It doesn't matter if it is the observer, the emitter, or both that are in motion. What matters is that the two objects are separating with respect to each other, causing the perceived stretching of the light's wavelength. When this kind of stretching occurs, it is called a red shift; the light's wavelength appears to be shifted into the redder portion of the light spectrum and the object itself appears redder. Similarly, when two objects are approaching each other, an emitted wavelength of light appears to shorten, and this shortening of wavelength is called a blue shift. The amount of the wavelength shift to the red or the blue depends upon the speed at which the objects are separating or approaching.

Elements such as hydrogen and helium, the principal components of stars, possess a sort of fingerprint. Each element emits light at known frequencies, called a light spectrum, that are slightly different for each element. In the 1920s, the astronomer Edwin Hubble observed the Doppler shift of stellar hydrogen and helium spectra with the 100-inch telescope at the Mount Wilson Observatory in California. He was the first to notice that these spectra are Doppler shifted into the red. That stars in distant galaxies are redshifting means that those galaxies are moving away from our own. Through his observations of the spectrum shifts of hundreds of galaxies, Hubble found that the farther away the galaxy, the faster it appears to be moving away from us. Astronomers today also measure the red shifts of quasars, extremely bright emitters of light whose origins are still a subject of debate. Some believe that when they are observing a quasar they are witnessing the birth of a galaxy, while others believe that they are observing the echoes of a much earlier time in the universe before galaxies formed. These quasars, from their red shifts, are observed to be moving rapidly away from us, and appear to be some of the most remote objects in the universe.

A cluster of stars observed through a prism reveals their different spectra. The spectra of older stars that have burned all their hydrogen have bright colored lines, while the spectra of younger stars have dark lines caused by elements in the stars' outer layers absorbing the emitted light.

The surprising discovery that all objects in space are moving away from us again seemed to put the Earth at the center of the universe. Since it did not seem appropriate to go back to the time before Galileo in our conception of the Earth's position, astronomers began searching for a different cause for this newly observed phenomenon.

The Mass of the Universe

Several centuries before, Isaac Newton had already begun to think about the extent of the universe. Newton surmised that if the universe extended infinitely, there would be no center, and no place around which gravity could attract all the matter of the universe. If the universe was *not* infinite, a center could be located, and gravity could pull all the matter together, causing the universe to collapse into one place. Even though Newton thought about the extent of the universe and questioned gravity's role in its fate, he would never have suspected that the universe is actually expanding!

The theories of Albert Einstein and his followers would provide a plausible explanation for the observed expansion of the universe. In the present-day formulation of the theory of general relativity, it is space itself that is expanding. According to this theory, the more space there is between the observer and an object, the more space there is to expand between the two and so the faster they would appear to be moving apart. Just as the seeds within a growing watermelon are carried farther away from each other by the expanding pink fruit, so galaxies are "carried" farther away from each other by the expanding space. In the growing watermelon, the seeds farthest apart will show the largest amount of movement away from each other, just as the most remote galaxies appear to be moving away from us the fastest.

If we conclude that space itself is expanding, the universe must have had its origin at a point back in time when it was smaller. At a moment sufficiently far back in time, the whole universe must have been a small point. For this chapter in our story, I shall limit the discussion to the future of the universe; I will leave the possible events of the distant past until Chapter 9.

Our new knowledge that space is expanding, taken with what we already know about mass and gravity, provokes a question: Does there exist sufficient mass in the universe for gravity to stop, and then reverse, the expansion, causing the collapse of the universe?

There are four possible scenarios. First, if there were no gravity, the universe would continue to expand at the same speed imparted to it at some very remote time, referred to as the big bang. Although we are certain that gravity does exist in our universe, it is still useful to contemplate a universe without the gravitational interaction between matter. If the universe were expanding without the benefit of gravity, then even though atoms would have formed because of electric attraction, these neutral atoms could not have clumped together. The gravitational interaction between even the smallest particles is, therefore, the key to our question about the total mass of the universe. Since we know that gravity does exist and that all particles with mass in the universe, regardless of their separations, are coupled to all other particles of matter in the universe, we know that gravitational attraction will slow the universe's expansion rate.

There are now left to us three possible fates for our universe. First, if there is not enough matter to exert a sufficiently strong gravitational force, the universe will expand forever, slowing down but never actually stopping. The second possible fate would take place in the case where there is "too much" matter. In that case, the expansion will stop, all particles of matter will fall into each other, and the universe itself will eventually collapse. Finally, we have the third, very special and critical case in which the amount of matter is neither large enough to ensure that the universe stops expanding nor small enough to permit the universe to continue expanding forever. In this case the expansion period will eventually come to an end, and we will have a stable, nonexpanding universe. This very unlikely scenario would require just the correct amount of matter or the operation of a principle of physics from the original creation that has not yet been foreseen by science. Since it seems that the fate of the universe is dependent on the mass in the universe, we would be able to predict its fate if we could measure that mass.

Experimental evidence supporting this picture of the universe was provided in the mid-1960s when two scientists inadvertently detected background radiation emanating equally from all directions of space. At Bell Laboratories in 1965, two engineer-scientists studying satellite communications, Robert W. Wilson and Arno Penzias, built a new and very sensitive radio receiver. While trying to understand a noise problem in

Electromagnetic radiation spans a huge range of energies making up the electromagnetic spectrum. At higher energies the frequency of the radiation—measured in Hertz, or cycles per second, abbreviated Hz—is also higher.

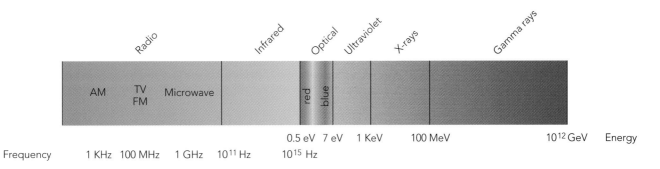

The observed spectrum of microwave radiation from the night sky. The spectrum has the characteristic shape of blackbody emission originating from an object that is cooling down. The peak of the spectrum indicates that it has a temperature of only 2.7 degrees Kelvin.

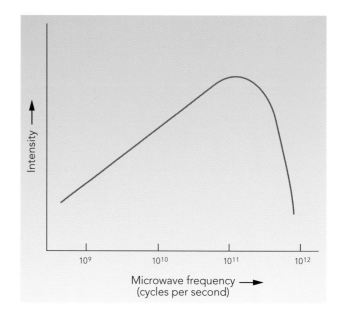

the microwave frequencies, they noticed a background noise from space in those frequencies that was uniform in all directions. This background noise had a recognizable origin in what is called blackbody radiation. When an object, such as a black cannonball, is left out in the Sun all day, it absorbs heat. It then releases the heat during the night while it cools. Since what we call heat is radiation in the infrared wavelengths, the cooling ball is emitting radiation from this part of the spectrum. If we watch these "thermal" emissions during the night, we observe the ball to emit radiation with a characteristic spectrum: the intensity rises slowly to a maximum as frequency increases, then falls sharply to zero. The microwave emission spectrum from the sky has a similar shape, signifying that the microwave noise had a similar origin through cooling. This microwave emission spectrum can be heard on a good FM radio: set one channel button to below 88 MHz, and a second one to above 108 MHz, but not near any station. You should hear just static on both channels. Then, if you rapidly switch between these two channels, you should hear a higher intensity of static on the higher-frequency channel. This is the low-frequency part of the microwave background emission spectrum. This effect might not be noticeable inside a big city, but comes through clearly in the bottoms of narrow mountain valleys.

The big bang theory predicts that the universe should be filled with a cosmic background radiation. The initial big bang would have emitted radiation that has spread out in all directions and cooled with time, currently striking Earth as microwave signals. Wilson and Penzias's measurements indicated that the average thermal temperature of this microwave background is the same as the temperature of deep space far away from our own sun—2.7 degrees Kelvin, or −454.8 degrees Fahrenheit, just a few degrees above absolute zero.

In the early 1990s, NASA launched a new scientific satellite called COBE, for *Cosmic Background Explorer*. COBE's job was to measure the uniformity of the cosmic background radiation in the blackbody microwave spectrum. This satellite had the ability to look at two parts of the sky simultaneously and compare their microwave temperatures with great precision. It was designed to look for non-uniformities in the microwave background that would indicate concentrations of matter. After completing its mission, COBE has found a uniformity to one part in 100,000, the sensitivity of the instruments. This uniformity lends further support to the big bang theory of the origin of the universe, a universe with no physical center or clumping of matter. The almost perfect uniformity also supports general relativity's hypothesis that space itself is expanding.

Missing Matter

Astronomers thought that they could estimate the total amount of mass in the universe by looking at the amount of luminous matter; with a value for the mass, they could predict the eventual fate of the universe. Locally—that is, in the region of our galaxy—we know reasonably well the distribution of all luminous matter, which our instruments detect directly, and even all nonluminous matter, which is invisible to our instruments. Since the gravitational pull of matter causes small shifts in the motion of visible stars, the motion of local stars betrays the whereabouts of the invisible matter and provides an estimate for the local density of all matter. For example, the motions of the orbiting planets would betray the amount of mass in the Sun's system to an observer at another star of somewhat close proximity. The mass measurement taken by the observer would not be the mass of the Sun itself, but the mass of the whole solar system because all the planets are gravitationally bound to the Sun and would thus be observed to move as one combined body. The ratio of local matter density to luminous matter can be used to estimate the mass of the rest of our galaxy as well as the mass of remote galaxies. From the luminosity of all known galaxies (corrected for the local invisible matter), an estimate of the mass in the universe has been deduced. This mass was found to be insufficient to cause the universe to collapse. One would need to find 10 to 100 times more invisible matter for there to be enough to cause such a collapse.

Because a galaxy is composed of individual stars attracted to each other by gravity, it does not rotate as a rigid body. It moves instead more like the liquid in a toilet bowl being flushed, with different velocities at various distances from the center of the spiraling vortex. Stars rotate around the central core at different speeds, and those nearer the center move faster than those farther out. Astronomers can measure a distant galaxy's speed of rotation by comparing the Doppler shift from one arm of that galaxy to the Doppler shift of its opposite arm. By making the comparison at different distances from the core of the galaxy, they can obtain a rotation curve, a measure of the speed of rotation as a function of the distance from the center. This rotation curve provides a means of estimating the mass density of the galaxy at any particular distance from the core.

Astronomers have measured the rotation curves of spiral galaxies that can be observed nearly edge-on from Earth. To their surprise, they find that the curves do not have the shape predicted from mass estimates of luminous matter. The velocity of stars and matter orbiting a galactic core should after a point decrease linearly with distance away from the core. Rotation velocities would be expected to slow down as the matter density decreases farther away from the center. It is clear from photographs of galaxies that luminous matter does fall off, but velocity measurements indicate that stars maintain a high rotation speed beyond this radius. It is as if there is much more matter away from the galactic core than we can observe.

A new question emerges: Can this dark matter, or whatever it is that seems to be at a far distance from galactic cores, have mass sufficient to change our estimate of the mass of the universe? If so, will the revised estimate change our prediction about the fate of the universe?

Although astronomers volunteer their own speculations as to where nonluminous dark matter may be hiding—in supermassive black holes or in numbers of brown dwarf stars scattered throughout the galaxies—particle physicists also have explanations to offer. As candidates for the missing matter they propose the possible existence of exotic particles such as magnetic monopoles, very weakly interacting massive particles, and many others beyond the scope of this book to describe. One candidate, however, is the neutrino.

As I shall discuss in Chapter 9, many billions and billions of neutrinos have been produced since the birth of the universe, more copiously at the beginning than now. Even if the neutrino's mass were only 22 eV/c^2—1/25,000 the mass of the electron—it would account for the missing mass of the universe; these massive neutrinos would eventually cause the universe to collapse.

The amount of matter in the core of a galaxy is extremely high. Since these core stars formed and ignited, they have produced a large number of neutrinos, which propagate outward. These neutrinos could have already traveled beyond the edge of the luminous matter in a galaxy, and if they have a nonzero mass, they would contribute to

OBSERVED ROTATION CURVE

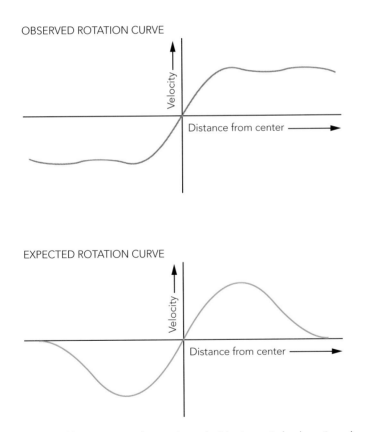

EXPECTED ROTATION CURVE

Astronomers are able to measure the rotation velocities in a spiral galaxy viewed nearly edge-on, like the galaxy shown at the top; from these velocities they obtain a rotation curve (bottom). However, this measured rotation curve generally does not match the expected rotation curve calculated from the distribution of luminous matter (middle). The obvious conclusion is that invisible dark matter is present.

the mass density there, explaining the observed rotation curves of the galaxies.

The early big bang produced such a tremendous number of neutrinos that they dominate neutrinos from all other sources. These big bang neutrinos are thought to be uniform throughout the whole galaxy. They therefore cannot be the dark matter "seen" in the rotation curves of galaxies, but can they account for enough mass to cause the collapse of the universe?

Determining the Neutrino Mass

In the same 1934 paper on beta decay in which Fermi solved the conundrum of the electron's energy distribution, he also correctly explained the effect that the neutrino mass would have on that distribution. An electron from a beta decay achieves the highest energy possible for it when all the mass lost in going from the initial nuclei to the final particles is converted to kinetic energy and given to the electron, rather than to the resulting nucleus or neutrino. This highest electron energy is called the electron's end-point energy, and it obviously depends upon the mass of the neutrino. If the neutrino has a zero mass, then the electron-energy distribution would gradually drop off to this end-point value; if the neutrino mass is not exactly zero but still small, then the distribution would approach this end-point energy slowly and then, just at the expected value, turn down sharply. If the neutrino's mass is large, the distribution would drop down fairly steeply at the highest possible electron energy with an almost sharp cutoff. All that is needed to determine the mass of the neutrino is a measurement of the shape of the electron's energy distribution at its highest, end-point value.

Although to the physicists of the time obtaining this measurement sounded fairly simple, it turned out to be difficult in practice. Electrons acquire an energy near the end-point value only when most of the energy goes directly into the electron, and the number of such beta decays is small. Typically the energy of the decay is shared between the electron and the neutrino. Contamination adds to the complexity of this measurement. Finding an uncontaminated sample of beta-decay elements, one that is free of elements that emit higher-energy particles, is extremely difficult. All of this, combined with the presence of ionizing particles that are distributed naturally in the air or the equipment, makes the task much more formidable. Even though the theoretical process through which the neutrino's mass could be measured had been understood since 1934, no measurement was made until 1952, when L. M. Langer and R. J. D. Moffat measured the neutrino mass as less than 250 eV/c² in the decay of tritium (^{3}H), a rare form of hydrogen having two neutrons in its nucleus in addition to the usual proton. Tritium decays into helium-3 (^{3}He), with its single neutron and two protons, releasing an electron and a neutrino:

$$^3H \rightarrow {}^3He + electron + neutrino$$

Although Langer and Moffat's result allowed the neutrino to have a small mass, it also implied that a zero mass was possible.

Real excitement came in 1980 when scientists participating in the Soviet beta-decay experiment at the Institute for Theoretical and Experimental Physics (ITEP) in Moscow performed a new and highly refined experiment. The ITEP physicists measured the mass of the neutrino in the same tritium-decay process as definitely nonzero, estimating its mass as greater than 14 eV/c^2 and less than 46 eV/c^2. Not only did this result excite physicists, but it was in the exact range astronomers needed in order to understand the big bang theory of the universe's expansion and the missing dark matter.

After the first definite measurement of the neutrino's mass, physicists elsewhere quickly set up new experiments to confirm the result. They soon realized, however, that the ITEP results were slightly flawed because the experimenters had used a frozen source for their tritium, which caused a distortion in the electron-energy distribution as the electrons left the dense, solid source. The new experiments were designed to avoid such a problem, in the hope of obtaining a more accurate value.

A group at the University of Zürich carried out one of the first series of experiments, and it still has the best measurements. This seems fitting, since Zürich was the home of Wolfgang Pauli, who, as we know, had first conceived of the neutrino. Using a gaseous source of tritium, the Zürich group accelerated electrons emitted from the decay by applying a precision electric field. Accelerating the emitted electrons gives

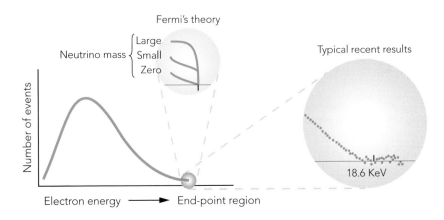

Fermi's theory predicts the shape of the electron-energy distribution in beta decay near the end point for different neutrino masses. The shape actually measured experimentally, shown at the right, best matches the zero-mass curve.

them a higher energy and produces a more accurate measurement of the electron energy. As in the ITEP experiment, the exact energy was measured by bending electrons in a magnetic field and recording their final position with a position-sensitive detector. In 1986, the first results gave a neutrino mass consistent with zero but not greater than 18 eV/c^2, a value substantially different from the ITEP's value of 35 eV/c^2. The University of Zürich's latest result, from 1990, gives a mass for the neutrino that is still consistent with zero and definitely less than 9 eV/c^2, or 1/60,000 the mass of the electron.

This result was disappointing to cosmologists. The sought-after missing mass in the universe could not be pinned on the neutrino unless a substantially greater number of neutrinos were produced during some phase of the early universe than theories suggested. Currently, however, astronomers and cosmologists are studying "inflationary" big bang models of the universe that better fit the Zürich experimental data. There is hope yet in some theories that the neutrino will be the answer to the missing-matter problem, especially new types of neutrinos that will be introduced later.

It should be mentioned that other accurate experiments have refuted ITEP's initial result. At the Los Alamos National Laboratory, the mass of the neutrino has been found to be less than 26.8 eV/c^2 — still consistent with a zero mass. Because none of these new experiments could confirm the ITEP result, it is no longer considered to be valid. Independent groups must confirm an experiment before its results can be accepted as scientific fact.

. . .

In 1934, Enrico Fermi explained how the mass of the neutrino could be measured without having to directly detect the neutrino itself. Nevertheless, no experiment designed to attempt this measurement has been sufficiently sensitive to measure the mass exactly, although such experiments have allowed limits to be placed on the value. Some excitement was created in 1980 when an experiment obtained a definite, nonzero mass, just sufficient to explain the missing matter in the universe, but it turned out to be the result of an error in the measurement technique.

Although the long-term fate of our universe is uncertain, a window into the future would be provided if a sufficient amount of "missing" dark matter were shown to exist. It seemed, at first, that the neutrino qualified as a candidate for this missing matter. The hopes raised by the neutrino mass obtained from the Soviet beta-decay experiment in 1980 provide a good example of how physicists expect knowledge about elementary particles to shed light on larger structures such as galaxies. Although today this excitement has lessened, physicists still believe that the existence of large amounts of small particles can influence the fate of the universe.

The current best limit on the neutrino's mass comes from the University of Zürich, which produced results consistent with a zero mass for the neutrino, but allows a mass of up to 9 eV/c^2. Further experiments with better resolution or more statistical data are needed to improve on this result. Although the Zürich limit eliminates the neutrino as a candidate for the missing dark matter in the universe under most common models, other exotic phenomena, such as speculative new elementary particles or black holes, still qualify. A neutrino telescope would offer a means of searching for such space exotica, as explained in Chapter 8, and the study of neutrinos helps illuminate events in the very early universe, as we shall see in Chapter 9.

An invisible photon, entering from above, collides with an atom, producing an electron-antielectron pair, whose trajectories are visible as the two spirals in this false-color bubble chamber image. These particles of negative and positive charge curve in opposite directions in a magnetic field. Understanding the theory behind these simpler interactions, governed by the electromagnetic force, led to more advanced theories that explained how the neutrino was related to the electron in beta decays. The long, straighter track is a recoiling atomic-shell electron.

CHAPTER 3

Where Have All the Right-Handed Neutrinos Gone!

Fermi's theory of beta decay correctly described all known experimental results of his time, but it did not seem to be a fundamental theory of physics. With four particles joined at one interaction point, it seemed to be more complicated than electromagnetic interactions with only three. Something more fundamental appeared to be at work in the beta decay of atoms, something that had not yet been discovered. Although physicists were about to take many experimental and theoretical steps that

Hideki Yukawa (right) with Werner Heisenberg in Geneva in 1958. Both physicists played essential roles in establishing the theoretical foundation of weak-force particle interactions.

would make for a better understanding of beta decay, the neutrino itself was about to become more perplexing than anyone could have imagined.

Fermi's four-point beta-decay theory was a great triumph in that it solved the thirty-year-old mystery of the beta-decay electron-energy spectrum. It was also an important step in the development of a more general theory describing all interactions mediated by the weak force. However, the theory would have to be modified before the current theory of weak interactions could evolve. In 1936, shortly after Fermi published his beta decay paper, a Japanese physicist by the name of Hideki Yukawa proposed a modification. Yukawa suggested that the weak force was similar in behavior to the electromagnetic force governing individual particle interactions at the quantum mechanical level.

Understanding Particle Interactions

According to the quantum mechanical theory of electromagnetic interactions, every particle that carries charge is continually emitting a cloud of "imaginary" photons in all directions. Most of these photons never encounter anything, but if one encounters another charged particle it can be absorbed. It is by the exchange of one of these photons that charged particles interact through the electromagnetic force: they will be attracted if of opposite charge and repelled if of like charge. This photon is a type of particle called a "virtual" particle, as opposed to a real particle, and has some special properties that we will learn about below. These photons can carry momentum, but are like ordinary photons of light only in being limited to the speed of light *c*. Ordinary photons do not propagate the electromagnetic force, so a bright light bulb will not be attracted to the object it is illuminating.

An analogy suggests the role of the virtual photon as mediator of the electromagnetic force. Imagine two children ice skating on a pond. They've picked up a ball and are tossing it back and forth. When one child throws the ball, he slides away from the direction in which he has thrown the ball—as required by the conservation of momentum. The child catching the ball will also slide backward (i.e., in the direction in which the ball was thrown), as also required by the conservation of momentum. As these children continue to play their tossing game, they drift farther apart with every toss. These actions are analogous to how two particles of the same charge exchange a virtual photon and how that exchange gives rise to electrostatic repulsion.

Suppose that, instead of playing a tossing game with a ball, the two children had a rope. If one child hands the rope to the other without letting go and if they each tug on the rope, they would be simultaneously pulled closer together. This action is analogous to the attractive force of two particles of opposite charge, which also exchange a virtual photon.

Virtual particles owe their existence to the famous Heisenberg uncertainty principle. From quantum mechanics we know that space and time at the atomic scale are different from how we perceive them at the large scale of visible objects. The rules governing the interactions of visible objects, referred to as *classical* mechanics, do not apply to the very microscopic interactions of elementary particles. On a sufficiently small scale, point particles behave as though they can be in two places at once. This property is predicted by the Heisenberg uncertainty principle, which is expressed mathematically as:

$$\Delta x \, \Delta p > h/2\pi$$

or alternatively

$$\Delta E \, \Delta t > h/2\pi$$

The delta symbol (Δ) in front of a variable means the uncertainty in that variable. The constant h is Planck's constant, 6.58×10^{-16} (eV seconds), a small number that limits the uncertainty in the product of position x with momentum p, or energy E with time t. The importance of Planck's constant, h, cannot be overstated because it is to quantum mechanics what the speed of light, c, is to relativity. The quantum interpretation of this equation is, in the upper version, that it is impossible to measure simultaneously the position of a particle and its momentum more accurately than $h/2\pi$ and, in the lower version, that it is impossible to measure simultaneously the energy of a particle and its lifetime. In particle interactions, these formula represent the extent over space and time that a given particle may have, and the extent of its energy or momentum, all limited by the uncertainty principle. Through the uncertainty principle, a particle can seem to jump from one place in space to another instantaneously, disregarding the conservation of energy constraints imposed by classical mechanics.

This disregard for mass to energy equivalence could be considered the defining characteristic of a virtual particle. Take, for example, a particle decay. If the mass difference between the final and initial particles is sufficient to liberate the requisite energy, then the particles created in the decay are not virtual and they live longer than the limit imposed by the Heisenberg uncertainty principle. However, when the decay does not liberate enough energy, that particle may still be created as a virtual particle. Its length of existence is limited by the uncertainty principle, but over the short time for which it exists the reaction does not need to satisfy the mass to energy conversion requirement, as a real particle produced in the decay would have to. A virtual particle thus makes a kind of trade-off: it is allowed to briefly violate standard conservation laws, but its own existence must be very brief. A virtual particle does carry momentum, but it does not have mass (neither do some real particles). So a virtual particle carries information from one point to another, but differently from a real particle.

In a sense a particle jumping from one place in space to another thanks to the uncertainty principle can be considered to exist in two places at once for a very limited time, again only within the limits of the uncertainty principle. This holds true for both real particles and virtual particles. The formula for the uncertainty principle gives us the largest distance a particle may perform such a virtual jump and not be required to obey the conservation of energy principle.

How far a virtual particle jumps is related to how long it exists. The uncertainty principle tells us that a virtual particle with an energy of 100 MeV can exist for 6.58×10^{-24} second, regardless of energy conservation. This is a very small amount of time. The maximum speed any particle can travel is the speed of light, 3×10^8 meters per second; multiply the speed of light by the particle's lifetime, and we find that the particle may virtually leap an incredibly small distance of 2×10^{-15} meters. A virtual particle of large mass can exist only over a small time or an equivalent distance. A particle of no mass, such as the photon that mediates the electromagnetic force, can propagate its force across the universe, although it can still travel no faster than the speed of light. If the photon had only a slight mass, its range would be highly limited, reducing the electromagnetic distance of interaction. All of these effects have their origin in the realm of quantum mechanics and do not apply to macroscopic particles.

Richard Feynman, an outstanding American physicist, created a technique for drawing pictures of electromagnetic interactions that would allow for theoretical calculations. The theory in question is quantum electrodynamics, which describes the interaction of charged particles, such as the electron, through the exchange of virtual photons, the propagator of the electromagnetic force. Through Feynman's pictorial descriptions, the details of the theory become assessable. Although the technique of drawing particle interactions was not new, Feynman refined it so that, when the Feynman rules of quantum electrodynamics were applied to all the graphic depictions of an interaction, the complete physical equation that described the interaction would be provided, allowing theorists to calculate measurable quantities.

The rules for correctly drawing Feynman diagrams in quantum electrodynamics are relatively easy to learn. Solid lines represent charged particles; a wavy line represents the photon, the propagator of the force. A diagram has time running along the horizontal axis and space along the vertical, although a common misconception is that both axes represent space dimensions, which is not technically correct.

A few examples will demonstrate how these diagrams work. A particle moving at a 45° angle on the graph is moving close to the speed of light, while a particle moving horizontally is traveling through time only—it is stationary in space. A particle moving vertically in a Feynman diagram travels instantaneously from one point in space to another and must be a virtual particle.

The simplest interaction to draw as a Feynman diagram is the annihilation of an electron and its positively charged antimatter counterpart, the positron. Where these two particles meet, they annihilate, producing a photon. The photon will travel forward through time briefly, but to conserve momentum this photon must be virtual in time, so it converts back into an electron-positron pair. However, if the electron-positron annihilation takes place near an atom, where momentum could be conserved, then the process can produce a real photon. Since the photon exists along the time axis, it does not act as the propagator of the electromagnetic force in this example. It is interesting to note that a positron moving forward in time can also be interpreted as an electron moving backward in time.

As we saw in Rutherford's alpha-scattering experiments, when the positive alpha particle approaches the positive, concentrated core of the nucleus, a repulsive force causes the particle to be deflected. We can show this type of interaction by drawing a virtual photon that vertically connects the two charged-particle lines. The same graph can be used to depict the interaction of two particles of opposite charge, except that the electric force would be attractive rather than repulsive. In this example, the photon is acting as the propagator of the electromagnetic force. Once all the diagrams relevant to a process or interaction have been

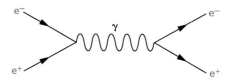

A Feynman diagram of an electron-positron annihilation. The diagram shows the initial production of a photon (γ) and the subsequent creation of an electron-positron pair from the photon's decay.

A Feynman diagram of Rutherford's alpha-scattering interactions. The scattering is caused by the positive alpha (α) particles' repulsion away from the concentrated core of positively charged protons in the nucleus.

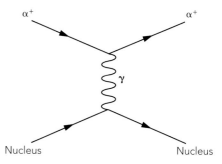

drawn, then any measurable quantity—such as scattering angle distribution or the distance of closest approach between the charged particles—can be calculated. Some interactions to be correctly described require more than one graph. For example, the third-order correction for the electromagnetic moments requires a mere 72 diagrams!

It is not unheard of that a theorist forgets a few possible diagrams in a high-order calculation, so that changes are required to his or her previously published results. In one instance, a theorist's spouse, after looking over a series of Feynman graphs, pointed out a few alternative ways of drawing the graphs. The missing graphs affected the result so much that another paper, this time with the correct calculations, had to be published.

The Weak-Force Propagator Particle

Yukawa understood that if the electromagnetic force acted through a special particle, then so could the weak force. He postulated the existence of a negatively charged particle that mediates the weak-force "charge" carried by two particles, much like the photon mediates the electromagnetic force between two electrically charged particles. Although Feynman did not create his diagrams to illustrate weak nuclear processes, the diagrams that carry his name have been extended to that force and the strong nuclear force. Yukawa's new particle allowed a beta decay to be rewritten as two three-point interactions joined together by his new force-propagator particle, called here the W, but sometimes called the intermediate vector boson, the symbol for which is a boldface W. This proposal of Yukawa's was a simplification compared to Fermi's four-point theory. Fewer particles are connected at each vertex (where the lines meet at a point), making it easier to write equations describing the interaction. Fermi quickly accepted the proposal because it made the vertexes look similar to those of photon interactions.

This new particle would help to explain why the weak force is 10,000 times weaker than the repulsion of the electromagnetic force. Heisenberg's uncertainty principle tells us that any massive particle can be created as a virtual particle, but its existence in that form is limited to a thin slice in time. If the W particle were to have a high mass, it could not be physically created in the decay process; it could exist only as a virtual particle, as presented in the Feynman diagram of beta decay. Note also that no time passes during the life of the particle, so its existence, even if its mass is large, does not violate the conservation principles. As a virtual particle with a brief existence, the W would propagate over exceedingly short distances only. Thus the distance over which it can function would be extremely limited and the force appears to be weak.

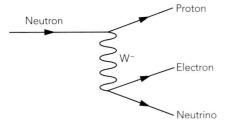

A Feynman diagram of normal beta decay through Yukawa's proposed W particle, the mediator of the weak force.

By late 1934, Frederic Joliot and Irene Joliot-Curie had discovered a new form of nuclear beta decay. In normal beta decay, a neutron decays into a less-massive proton plus an electron and a neutrino. In this new decay, a proton—in this case from the sodium-22 (^{22}Na) nucleus—decays into a neutron, creating a neon-22 (^{22}Ne) nucleus and in the process emitting a positron-neutrino pair coupled to the W. Energy is conserved: the sodium-22 nucleus is actually heavier than the neon-22 nucleus and the mass of the emitted positron taken together.

This new decay process fit Yukawa's proposed theory well. It requires only that the W particle come in a positive charge as well as in the negative form in which it is found during normal beta decay. Another significant feature of Joliot and Joliot-Curie's new beta decay was that the decay occurs even if the initial particle, in this case the proton, is lighter than the final particles, the neutron and positron. The mass of importance in this interaction is the mass of the initial nucleus, which must be greater than that of the final nucleus and the positron together.

Luis W. Alvares and Oskar Klein discovered a third beta-decay process in 1938. They found that a proton was able to capture an electron from the atom's own atomic shell. The proton is transformed into a neutron, while simultaneously emitting a neutrino. A distinctive feature of this decay is that the element changes into an element of lower charge rather than higher. Although this process does not occur in every kind of nucleus, it is a common type of radioactive decay. The new element's nucleus is usually unstable and will decay through any of the three known radioactive processes, but most often through beta or gamma decays.

We now have three ways for a nucleus to undergo beta decay; all of them can be explained by the weak-force propagator particle devised by Yukawa, and all of them involve the emission of a neutrino. However, in order to explicitly prove Yukawa's new theory, physicists had to observe the production of real W particles. Since the theory and experimental results did not give the exact magnitude of weak interactions, it was not certain what mass the new weak propagator particle would have. The actual production of this particle, as well as the measurement of its mass, would have to wait fifty years (but we will have to wait only until Chapter 7).

Parity Violation in Weak Decays

In Chapter 1 we saw that the conservation of energy principle is a presumed law of nature. When confronted with the unusual electron-energy distribution of beta decay, physicists had to contemplate the possibility that this law was not valid at the subatomic particle level for every event. In the end this solution to the problem was not necessary,

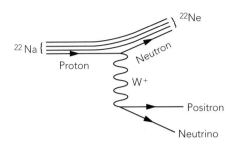

A Feynman diagram of a proton decaying; this decay requires the existence of a W particle with positive charge.

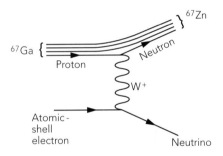

A Feynman diagram of the type of beta decay that proceeds through the capture of an electron.

and the energy conservation principle still stands today. From this experience physicists acquired a great appreciation for searching out new laws of physics to guide them into the future. One way to do this is to search out symmetries.

Nature has many symmetries—for example, a tree leaf or the human body is identical on either side of a line drawn down the center. In physics a symmetry is a property that can be inverted without changing the physics laws ruling an interaction. The easiest example of such a property is electric charge. It is well known that the laws governing electromagnetic interactions remain the same if all particles in an interaction change polarity, positive becoming negative and negative becoming positive. After such a charge reversal, abbreviated by the capital letter C, all particles attract or repel one another identically as if nothing had actually changed. In all known electromagnetic interactions, a charge reversal is known to produce the same physical results.

A similar law of physics exists for time reversal, abbreviated by the capital letter T. This symmetry principle tells us that at the microscopic level at which individual particles interact, any particle interaction that occurs forward in time can also occur backward, as if the event were seen in reverse. A subtle warning should always be kept in mind when thinking about time-reversal transformations: even though the reverse process is allowed, it might not be plausible. This is true at the particle level; at the macroscopic level of ordinary objects, such time reversals may be forbidden outright.

Imagine a home video of a vase falling from a table and shattering. If that video were run backward, the pieces of the vase would jump off the floor and reassemble onto the table. Although it is conceivable that the pieces of the vase on the floor could be imparted with just the right forces to make them jump back up the way they fell apart, the chances of that occurring are negligibly small, so the reassembly of the vase is not a plausible occurrence. At the macroscopic level of a falling vase, the pieces of the vase could not reassemble themselves without violating the laws of thermodynamics, but at the individual particle level, all processes that go forward in time should also be able to go in the reverse direction.

Parity conservation, abbreviated by the capital letter P, has also been presumed to be a law of physics. Simply put, a phenomenon observed in the reflection of a mirror obeys the same physical laws as the identical phenomenon observed directly, without a mirror. For gravitational and electromagnetic interactions, this law has been extensively studied and appears to always hold true. Until 1956, physicists also assumed it held true for weak interactions. They were to be surprised when they found that beta decays violated this presumed parity symmetry law, something unexpected to many scientists of the time.

To illustrate parity: an x-axis and y-axis are drawn as usual, with the x-axis appearing horizontally on a piece of paper and the y-axis appearing vertically. What is called a right-handed coordinate system would be created if you were to draw the z-axis so that it begins at the origin and appears to come out of the paper toward you. By reversing the direction of one of the axes, for example by making the z-axis appear to go away from you, you instead create a left-handed coordinate system. Going from a right-handed system to a left-handed system is equivalent to looking at a reflection in a mirror. Before 1956, it was commonly assumed that all laws of physics were identical in either right-handed or left-handed systems; that is, parity was conserved when one system was transformed into another.

Two young theorists, T. D. Lee and C. N. Yang, challenged the established physics community in 1956 by pointing out that there was no experimental evidence proving that parity was actually conserved in weak interactions such as beta decay. The difficulty these theorists and others had with parity was that, even if no experiment showed that parity could be violated, it would be extremely difficult to prove experimentally that parity is always conserved. It seemed to Lee and Yang that experiments designed to dig deeper into the nature of parity at least deserved attention before physicists jumped to any conclusions. They suggested many experimental tests, some of which were performed, with surprising results.

These experiments focused on the property of spin. Many particles spin: the Earth spins on its axis causing our 24-hour cycle of day and night; a rocket launched from Earth has a spinning, massive gyroscope to keep it stable during the booster phase of its flight. In nuclear physics, a particle's spin about its direction of motion has a special meaning. Using the right hand, point the thumb and curl the fingers; suppose that the thumb points along the particle's direction of motion. If the particle's direction of rotation matches the fingers' direction of curl, then that particle is said to have right-handed spin. If it does not, then the particle is said to have left-handed spin. There is one complication in this simple picture that must be considered. A massless particle has to have its axis of rotation oriented along the particle's direction of motion, but an elementary particle with mass does not. A special word, "helicity," is used to denote the quantity of spin about the direction of motion. For a massless particle this quantity is the total spin, whereas in the case of a particle with mass the total spin could be different from the measured helicity.

(As an aside, it's interesting to note that spin provided another mechanism by which physicists could substantiate the existence of the neutrino. At the elementary particle level, it is known that spin comes in discreet amounts just like charge, but unlike charge spin comes in full-

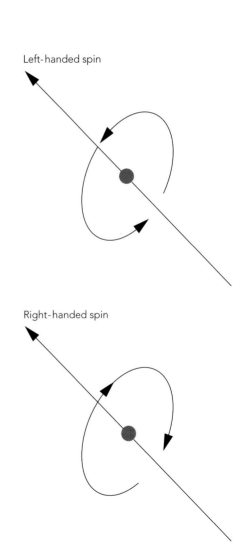

Left-handed spin

Right-handed spin

The spin of a particle around its direction of motion for left- and right-handed particles.

A spinning particle with mass can have its total spin not aligned with the direction of motion. In this case, the component of the spin that is around the direction of motion is called the helicity. For a massless particle, the total spin must be along or opposite the direction of motion, and it too is called the helicity.

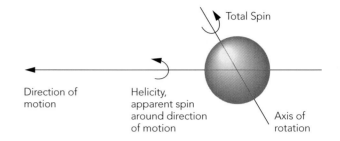

and half-integer increments of $h/2\pi$. In beta decay, the sum of the spins of the resulting particles must equal the spin of the initial particle; otherwise angular momentum is not conserved. Take the first type of beta decay as an example. The initial neutron has half-integer spin, as does the proton that results. However, the emission of an electron, also a particle of half-integer spin, must also be taken into account, making the before and after spins unequal. If the electron were the only particle emitted in beta decay, angular momentum would not be conserved. Therefore, to balance the half-integer spin of the electron, either another invisible particle of half-integer spin but opposite handedness must accompany the electron in its emission (i.e., the neutrino), or we must dismiss the momentum conservation law. Although this is a powerful argument in favor of the neutrino, the direct detection of the particle would still be most helpful.)

Electrons are emitted in one preferential direction from a frozen cobalt source. This preference was evidence for the violation of parity.

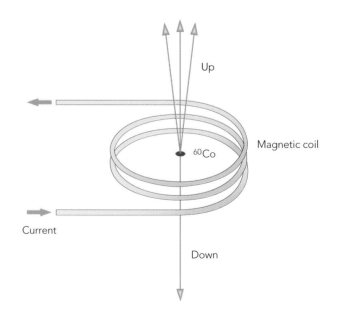

The first experiment to look for parity violation was performed at Columbia University in New York by Chien-Shiung Wu, with the help of the National Bureau of Standards. For their source of particles, she and her associates used frozen cobalt-60, which emits an electron and a neutrino through normal beta decay. They selected cobalt for use in this experiment because, owing to its particular atomic shell electron configuration, its nuclear magnetic field can be easily aligned in the same direction as that of an externally applied magnetic field. Aligning the nuclear fields would make all the cobalt atoms spin in the same direction. By freezing the cobalt source almost to absolute zero, they eliminated the thermal motion of the atoms, making the spins of the atoms easier to align with the applied magnetic field.

Once all the cobalt atoms were aligned to spin in the same direction, they all had the same identical coordinate systems relative to each other. It then became possible to see whether beta decay has a preferential direction in this coordinate system.

Wu's group had positioned the applied magnetic field so that an outgoing electron could travel in only one of two possible directions. If

Chien-Shiung Wu at the controls of an experiment.

Experimental evidence that demonstrates parity violation: a sketch of the results from Wu's experiment shows an asymmetry in the direction of electron emission from the beta decay of polarized cobalt atoms. The experiment was done twice with opposite magnetic-field settings; but the detector was left in the same place, where it was able to record only the electrons being emitted in one of the two directions. Thus the experiment observed many more neutrinos than average at one setting (down) and many fewer at the other setting (up). The asymmetry was seen to go away as the frozen cobalt source was permitted to warm up.

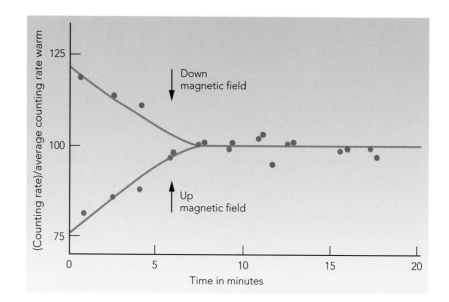

parity was conserved in this decay, then an equal number of electrons would emerge in either direction. But when they applied a magnetic field to align the spins of the initial cobalt-60 atoms, the electrons came out preferentially in one direction. When the magnetic field was reversed, the preferential direction changed. The preference for a single direction demonstrated that parity was not conserved in weak decays. The physicists also checked what happened when the cobalt source was allowed to warm up: the thermal motion of the atoms made them impossible to magnetize, and the atoms' spin alignment was lost, as was the preferential direction of the electron emission. This result demonstrated that the experiment had no systematic bias in the measurement.

This experiment showed that beta decays, a weak-force process, were emitting neutrinos that were preferentially left-handed. What it could not show explicitly was whether beta-decay neutrinos were always left-handed. Its failure to do so was explainable by the fact that even a strong magnetic field applied to a source frozen at absolute zero could not align the spins of all the cobalt atoms perfectly. Since any right-handed neutrinos emitted from aligned atoms could not be distinguished from left-handed neutrinos emitted from nonaligned atoms, an absolute result with this experiment was impossible.

After Wu and others had shown that weak decays violate parity conservation, it was proposed that charge and parity together might be conserved, a symmetry called CP conservation. If valid, the new principle would mean that when both parity and charge reversal are applied simultaneously, the laws of nature remain unaffected. To achieve charge reversal, one would have to make anticobalt-60, to use in place of cobalt-60, and reverse the magnetic field. If this anticobalt-60 were put into the same experiment and looked at as a reflection in a mirror, to simulate parity inversion of space, then everything should appear as if the source of the beta decays was regular cobalt-60. However, if parity reversal alone were applied, then the decay of anticobalt-60 would not look the same as the decay of ordinary cobalt-60. The identical outcome of an experiment when both charge and parity reversal are applied is referred to as the CP invariance principle.

At about the same time that Wu's paper appeared, other groups announced the discovery of parity violation in a different kind of weak decay (which will be introduced in Chapter 5). One of these papers, produced by a University of Chicago group led by Valentine Telegdi, was delayed in its publication when a death in Telegdi's family and his desire to refine the details of the presentation kept him from submitting it. This delay proved to be costly to the Chicago group.

Most scientists are motivated by the desire to make discoveries. The first person or group who publishes their work in a reputable scientific journal receives credit for new results, along with any patent rights that follow. For example, when Alexander Graham Bell submitted his patent request for the telephone only hours before his competitor, his promptness ensured him the monetary rewards that come with being the first to claim an invention. In order to help determine who will get the credit for a new result, all professional societies, such as the American Physical Society or the European Physical Society, publish the date on which the editors of a journal receive a paper on the first page of each article, which also includes the title and a list of authors. Each paper submitted for publication is reviewed by one or two scientists who judge whether the article is well founded and fit for publication. Physicists are normally reluctant to publish controversial new results because it puts their professional reputation on the line; if they are proven wrong, it would naturally be embarrassing. In general, the first scientist willing to risk his or her professional reputation by publishing a new result gets most of the credit. The publication procedure is just as essential a part of the scientific method as confirming an experimental result.

Because the Chicago group's paper was published after Wu's paper, they did not receive credit for the discovery of parity violation. Even though they were unhappy to have missed the glory, they do have the honor of confirming the results.

Wolfgang Pauli scolded the two young Chinese-American physicists who had taken the risky step of suggesting that perhaps parity was not conserved in weak interactions, joking that "the only thing new in their paper was a little bit of courage." Pauli, a conservative theorist, was convinced that symmetries of nature should be used to guide theoretical physics. He could not accept that "God was only a weak left-handed player." However, the Columbia University experiment and many others ultimately vindicated Lee and Yang.

As all of this was going on, Pauli, back in Zürich, was scheduled to give the Faraday lectures on physics. On the day of the lecture Pauli received preliminary papers from all the experimental groups showing that parity was indeed violated in weak decays. Charles Enz, Pauli's assistant, said that the paper by Telegdi's University of Chicago group was the straw that "broke" Pauli and convinced him to change his mind. In many ways Pauli had been Telegdi's mentor during Telegdi's years of study at ETH Zürich as a young man, and Pauli's great respect for his former student may have played a role in convincing him. Moreover, like any theorist, Pauli had to bow to overwhelming experimental evidence. That night, he gave the Faraday lectures, having properly changed his text to conform to the experimental proof of parity violation. Pauli even went out of his way to compliment Lee and Yang for their leading role, and his acknowledgment greatly helped them to receive the Nobel Prize in physics for their efforts.

The existence of parity violation in weak interactions suggested that the neutrinos of beta decay are only left-handed, but to address any reasonable doubts physicists still wanted to measure the phenomenon directly.

Directly Measuring the Helicity of the Neutrino

In the 1950s a group of physicists designed a beautiful and elegant experiment to measure the spin handedness along the direction of motion for the neutrino, the quantity given the name "helicity." It was one of the last great experiments in particle physics that could be done on a tabletop. That this quantity could be measured without directly detecting the neutrino is amazing. For the sake of completeness the following description is slightly technical, but a full understanding of it is not necessary to enjoy the story of the neutrino in the following chapters.

To directly measure the helicity of the neutrino, physicists needed to find a radioactive source that would have three special properties: it would decay by electron capture into a nucleus in an excited state plus a neutrino; the initial nucleus would have an orbital angular-momentum spin of zero; and the nucleus of the decay product would have an orbital angular-momentum spin of one (orbital angular-momentum spin is like the orbit of planets around the Sun, while spin is more like the rotation of a planet on its axis). Using a source with these properties would simplify the measurement. In the late 1950s, the only such source known was an isotope of europium, ^{152}Eu. Its special properties of orbital spin had been discovered by a group led by Maurice Goldhaber of Brookhaven National Laboratory, the group that would also perform the first direct measurement of the neutrino helicity. They were in an excellent position to do so because they were the only group who knew about these properties of ^{152}Eu. By the time their paper describing the decay properties of europium was published, they were well on their way to completing the helicity measurement.

The ^{152}Eu source undergoes the type of beta decay first found by Alvares and Klein. A ^{152}Eu nucleus captures an atomic shell electron, changing itself into a nucleus of the element samarium (^{152}Sm*) and emitting a neutrino (the raised star denotes an excited state of an isotope, which decays to its ground state by emitting a gamma ray). The nucleus recoils away from the neutrino, conserving momentum, and since both these decay products have the same helicity, a measurement of the nucleus's helicity would also provide the neutrino's. Furthermore, the excited nucleus decays to its ground state by emitting a gamma ray, so the nucleus's helicity may be measured from the observed polarization of the forward-emitted gamma rays. The emitted neutrino escapes undetected and is actually never recorded in this whole experiment.

Special detectors can record gamma rays, but the problem Goldhaber faced was to measure their polarization and eliminate background gamma rays. Since the gamma ray is emitted opposite the neutrino, it has a slightly higher energy than the 960 keV typical of a gamma ray from the decay of ^{152}Sm* at rest. This is fortunate, for it means that pure ^{152}Sm will be able to reabsorb and quickly reemit gamma rays that are emitted by an atom identical to itself, in a process called resonant scattering. The advantage is that resonant scattering acts like a kind of filter that only scatters these gamma rays and not others of the wrong energy. Two samples of ^{152}Sm adroitly positioned can scatter these gamma rays into an area shielded from unwanted radiation, helping to reduce the background noise created by other radioactive processes in the vicinity of the experiment.

As the sketch of the apparatus shows, the source is placed in a magnetic field. The gamma rays are emitted in all directions, but the rays emitted toward the detector are those that will be analyzed; the neutrinos emerge in the opposite direction, undetected. A cone of lead is inserted between the source and the detector, solely to shield the sensitive detector from direct gamma rays. By measuring the gamma-ray polarization, which indirectly yields the helicity of the unobserved neutrino, Goldhaber's team was able to make a detailed measurement of the neutrino's internal spin properties. The beauty of this experiment is that the physicists measured the neutrino helicity without ever directly detecting the neutrino itself.

If the gamma rays are polarized in the same direction as the magnetic field, then electrons of iron in the magnet are more likely to absorb the gamma rays. Gamma rays of one polarity would be transmitted and those of the other polarity would not be. Therefore gamma-ray transmission is greater for left-handed polarized gammas than for right-

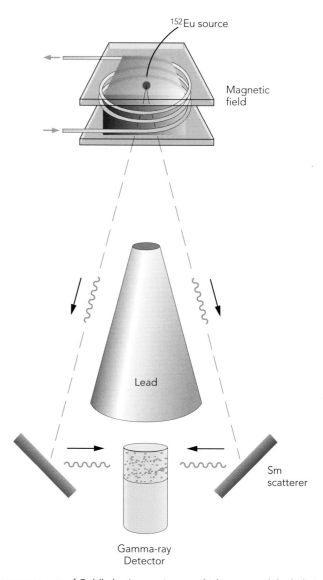

The major components of Goldhaber's experiment, which measured the helicity of the neutrino by measuring the polarization of gamma rays emitted along with the neutrino in the beta decay of ^{152}Eu.

handed ones. Changing the direction of the magnetic field, which reverses this transmission preference, allows for a comparison of left-handed and right-handed polarized gammas. The relative difference between number of gamma rays recorded at the two magnetic field settings gives the preference for left- or right-handed gamma rays, while the background common to both does not affect this measurement because the physicists looked only at the difference between the two magnetic settings. When the magnetic field was set up to absorb right-handed gamma rays, plenty of left-handed rays were present to set off the detectors. But when the magnetic field was set up to absorb left-handed gamma rays, the detectors fell eerily silent except for a small background identical to when there was no source at all. Results from this experiment show that the neutrino is 100% left-handed and no right-handed neutrinos existed at all from the beta decay.

So where have all the right-handed neutrinos gone? Essentially, we learned from this experiment that the emitted neutrino of beta decay exists in only one helicity state, and the other is excluded from occurring. No other experiment thus far has ever seen any but this one helicity state for the neutrino. In weak interactions, the electrons that accompany neutrinos also have only one allowable helicity. In quantum electromagnetic theory, in contrast, when a photon converts into an electron-positron pair, the electron has right-handed spin 50% of the time, and left-handed spin 50% of the time. In other words, both spin states are allowed. Why weak interactions produce only left-handed neutrinos while electromagnetic interactions allow for both right- and left-handedness was a perplexing problem that would not be quickly understood.

. . .

In the 1950s it was generally believed that certain symmetries in nature can be reversed without changing the physical nature of an interaction. One example that is still known to hold true is the charge symmetry observed in electromagnetic interactions. Changing the sign of all charged particles in an interaction does not affect its outcome, provided that all of the participating particles are reversed simultaneously. Parity, the symmetry of nature observed by looking at an interaction in a mirror, was also believed to be fundamental symmetry, but this belief was not to last. When studied experimentally, weak interactions were found to violate parity conservation.

As the existence of parity violation in weak interactions suggested, the neutrino has only left-handed spin about its direction of motion. Every other particle in nature may have either right-handed or left-handed helicity, so the neutrino is unique in this respect. Why right-handed neutrinos do not exist is a mystery that would not be fully understood until the late 1970s. Meanwhile, this new complexity presented a perplexing problem for physicists trying to understand the new force of weak interactions.

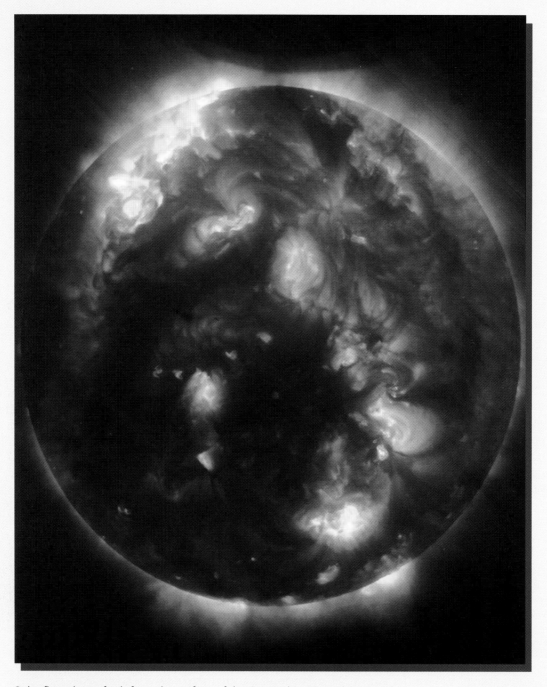

Solar flares burst forth from the surface of the Sun in this X-ray photograph, while the bright patches correspond to the dark sunspots seen in visible light. The X-rays and photons emitted from the surface acquire their energy from nuclear reactions in the Sun's core. These same reactions emit neutrinos that provide clues to conditions deep in the Sun's interior.

CHAPTER 4

Directly Observing the Neutrino

I n this chapter we will finally see the first direct detection of the neutrino. Observing the neutrino was to be no small feat; it would require the development of an immense source of neutrinos, such a source as would become possible only with the invention of nuclear reactors. After physicists had developed several techniques to directly observe neutrinos, scientists would turn these new tools to the study of neutrinos from the core of our Sun.

The Sun is a large furnace, fueled by nuclear reactions that generate heat, producing the light we see. These same nuclear reactions produce neutrinos, and if these particles could be detected then we could see directly into the core of the Sun itself. But once physicists and chemists had designed instruments to detect these solar neutrinos, they found fewer than current solar models predicted. Did this lack of neutrinos mean that our understanding of nuclear physics and the life cycle of the Sun was slightly flawed? Could the Sun's ultimate fading even be close at hand? I will discuss these questions in light of the most recent data, data that will also allow us to explore the core of the Sun itself.

Nuclear Reactions

Even though neutrinos are produced in every beta decay, they are such weakly interacting particles that the chances of stopping and observing a neutrino are extremely small. Because neutrinos interact only through the weak nuclear force, they have the smallest known cross-section of any particle, which simply means that the probability of their interacting with matter is extremely low. No detector exists that can record the passage of a neutrino. Instead, physicists have to rely on stopping a neutrino and converting it to other particles that can be observed.

It is not easy to stop a neutrino. Their chances of interacting are so small that, as noted earlier, a neutrino can typically pass unhindered through a lead barrier several hundred light-years thick. But if there are a sufficiently large number of these particles, a few will be stopped. Physicists hoping to observe one of these elusive particles needed a tremendous source of neutrinos to have a good chance of stopping one. Enrico Fermi supervised the development of just such an immense source during his work for the Manhattan Project, which the United States government had commissioned to develop the nuclear bomb. Before the bomb could be developed, Fermi's team would have to create a self-sustained nuclear reactor able to generate controlled nuclear reactions. These reactions would provide a powerful source of neutrinos, and with this source physicists would eventually be able to obtain solid evidence for the existence of the neutrino.

The story leading up to the harnessing of nuclear reactions, whether in a self-sustaining reactor or in a nuclear explosion, goes back to 1934 and experiments that Enrico Fermi carried out in Italy. After starting out as a theorist, Fermi had taken up experimental physics and accepted a position in a laboratory in Rome as an untenured professor of physics. Rutherford congratulated Fermi for his successful escape from the realm of theoretical physics, in a letter reprinted on the opposite page. It was Fermi's dual success as both an experimentalist and theoretician that, in some scientists' opinions, makes him the greatest physicist of this century.

Cavendish Laboratory,
Cambridge

23rd April, 1934

Dear Fermi,

I have to thank you for your kindness in sending me an account of your recent experiments in causing temporary radioactivity in a number of elements by means of neutrons. Your results are of great interest, and no doubt later we shall be able to obtain more information as to the actual mechanism of such transformations. It is by no means clear that in all cases the process is as simple as appears to be the case in the observations of the Joliots.

I congratulate you on your successful escape from the sphere of theoretical physics! You seem to have struck a good line to start with. You may be interested to hear that Professor Dirac also is doing some experiments. This seems to be a good augury for the future of theoretical physics!

Congratulations and best wishes,
Yours sincerely,

Rutherford

Send me along your publications on this question.

Fermi began his study of nuclear reactions with simple experiments. He bombarded heavy elements with neutrons to create temporary, artificial radioactivity in experiments that would become crucial to Fermi's later work on nuclear reactors and the bomb. At the time, no one understood the exact mechanism by which this temporary radioactivity was created, but it was eventually learned that the mechanism for nuclear reactions, whether in a continuous energy source or a bomb, was the release of the binding energy that holds particles together in the nucleus. The method is to take large nuclei, for example of elements such as uranium or plutonium, and separate them into two or more smaller

Temporary radioactivity is created when a reaction with a neutron splits apart a heavy nucleus, forming two smaller nuclei and releasing additional neutrons. These newly formed neutrons can then induce fission in other heavy nuclei, starting a chain reaction.

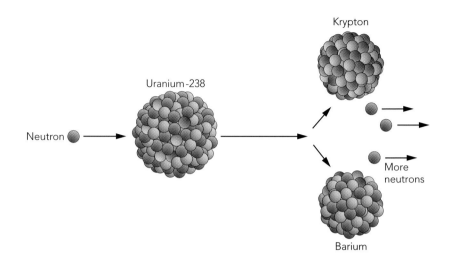

parts, a process called fission. As the pieces separate, the binding energy that has been holding them together is released. Fermi's 1934 experiments with the neutron bombardment of heavy elements achieved the first man-made fission reactions.

The other type of nuclear reaction, nuclear fusion, occurs when two light nuclei are pushed together to form one new nucleus. This type of nuclear reaction also results in a release of energy. Elements lighter than iron will not produce any net energy when they are split into smaller fragments; the binding force is so great that more energy must be provided to separate the parts than will be released. Elements heavier than iron, especially extremely heavy unstable elements, are easier to break apart, releasing much greater energy than is released in any chemical process. However, these heavy elements with a large amount of electric charge cannot produce sufficient fusion energy when pushed together. The large mutual repulsion to be overcome is much greater than the mutual repulsion of light nuclei; more energy than would result from fusion itself must be given to the heavy elements before they can come close enough to bind. Elements lighter than iron will release energy when fused together since the repulsion is easier to overcome. Iron is the break-even point: very little energy is released when an iron nucleus is broken apart, or when two small nuclei are brought together to form iron, relative to the amount of energy needed to cause the reaction.

It was Leo Szilard, a Hungarian quantum theorist, who in 1939 first recognized the military application of the work on nuclear reactions. Szilard encouraged Albert Einstein to write a letter to President Roosevelt suggesting the creation of a project to develop a nuclear bomb. Fermi and Szilard felt that only a well-known scientific figure could convince the United States government to pursue such a large project. Einstein wrote such a letter on August 2, 1939; it is reprinted on the next page. After subsequent meetings with the physicists, President Roosevelt made the decision to establish the Manhattan Project.

The project's first step was the construction of a self-sustaining nuclear chain reactor. For security reasons, it was necessary to move the project away from its original base at Columbia University in New York City. Its new home at the University of Chicago was considered more secure because of its geographic location in the interior of the United States and because it was located in a Navy port city with a military airport. Enrico Fermi, the project leader, moved his staff and family from New York to Chicago. In 1942 construction of the first Chicago atomic pile, called CP-1, started in the racket courts under the west stands of the Stagg Field football stadium. Project members assembled this pile from bricks of graphite, blocks of heavy elements rich in uranium, and cadmium rods. They placed the uranium-rich blocks in the center of the pile and laid cadmium rods between the layers of blocks. The cadmium hindered the propagation of neutrons, so the scientists would remove and insert these rods as necessary to control the speed of the reaction. Because the war effort had created a manpower shortage, the University of Chicago football team had to be enlisted to carry the heavy blocks. Project members hand-positioned all the supporting blocks, graphite blocks, and uranium-rich blocks into place.

On December 2, 1942, the pile was put into operation. A scientist partially withdrew a cadmium control rod, and the reaction started. A Geiger counter placed inside the pile showed the radioactive decays increasing in intensity as the control rod was further withdrawn. At first the meter on the counter indicated that the reaction rates being achieved were not self-sustaining; instead, they were leveling off at fixed, stable points, different for each control rod position. These leveled-off rates would be sufficient to generate energy in small amounts, but they were not conclusive evidence of the kind of self-sustaining reactions needed to produce a nuclear explosion. Slowly the scientist withdrew the cadmium rod even farther as the observers watched the increased activity. Finally, after several attempts, they noted a sharp, exponential rise in intensity with no sign of leveling off: a self-sustaining reaction had been achieved. The control rod was reinserted, and the pile's radioactivity level quickly dropped, indicating that the reactor was shut down.

This triumph marked a turning point in the Manhattan Project, for it was now clear that an explosive device could be manufactured. The results of the then-secret project are now well known. Three nuclear bombs were designed and assembled under the leadership of Robert Oppenheimer. After a successful test detonation in the Nevada desert,

Albert Einstein
Old Grove Rd.
Nassau Point
Peconic, Long Island
August 2nd, 1939

F.D. Roosevelt,
President of the United States,
White House
Washington, D.C.

Sir:

Some recent work by E. Fermi and L. Szilard, which has been communicated to me in manuscript, leads me to expect that the element uranium may be turned into a new and important source of energy in the immediate future. Certain aspects of the situation which has arisen seem to call for watchfulness and, if necessary, quick action on the part of the Administration. I believe therefore that it is my duty to bring to your attention the following facts and recommendations:

In the course of the last four months it has been made probable—through the work of Joliot in France as well as Fermi and Szilard in America—that it may become possible to set up a nuclear chain reaction in a large mass of uranium, by which vast amounts of power and large quantities of new radium-like elements would be generated. Now it appears almost certain that this could be achieved in the immediate future.

This new phenomenon would also lead to the construction of bombs, and it is conceivable—though much less certain—that extremely powerful bombs of a new type may thus be constructed. A single bomb of this type, carried by boat and exploded in a port, might very well destroy the whole port together with some of the surrounding territory. However, such bombs might very well prove to be too heavy for transportation by air.

The United States has only very poor ores of uranium in moderate quantities. There is some good ore in Canada and the former Czechoslovakia, while the most important source of uranium is Belgian Congo.

In view of this situation you may think it desirable to have some permanent contact maintained between the Administration and the group of physicists working on chain reactions in America. One possible way of achieving this might be for you to entrust with this task a person who has your confidence and who could perhaps serve in an inofficial capacity. His task might comprise the following:

a) to approach Government Departments, keep them informed of the further development, and put forward recommendations for Government action, giving particular attention to the problem of securing a supply of uranium ore for the United States;

b) to speed up the experimental work, which is at present being carried on within the limits of the budgets of University laboratories, by providing funds, if such funds be required, through his contacts with private persons who are willing to make contributions for this cause, and perhaps also by obtaining the co-operation of industrial laboratories which have the necessary equipment.

I understand that Germany has actually stopped the sale of uranium from the Czechoslovakian mines which she has taken over. That she should have taken such early action might perhaps be understood on the ground that the son of the German Under-Secretary of State, von Weizsäcker, is attached to the Kaiser-Wilhelm-Institut in Berlin where some of the American work on uranium is now being repeated.

Yours very truly,

Albert Einstein

(Albert Einstein)

the United States dropped two bombs on Japan, and World War II came to a speedy end in the Pacific.

Scientists usually work in an atmosphere of openness, relatively oblivious of political boundaries. Each achievement is published for others to read. This openness can sometimes lead to problems; for example, during the Cold War, a Soviet physicist was oftentimes suspected

A contemporary sketch of the first nuclear pile reactor, called CP-1, built during World War II under the bleacher stands of Stagg Field on the University of Chicago campus.

of betraying his country when scientists in the United States quoted the scientific results of his work in their papers. But even during the many years of the Cold War, important scientists from the countries of the so-called Western alliance and from behind the Iron Curtain attended joint physics meetings and freely exchanged their most recent results.

The team of scientists from the United States, Great Britain, and Canada who worked on the Manhattan project included many scientists whose peculiarities of background and behavior led to suspicions of treason. Most such suspicions turned out to be highly unfounded, although at least one physicist of the Manhattan Project, Karl Fuchs, was caught and jailed after admitting to spying for the Soviet Union. At the beginning of the Manhattan project, it was Fermi and his team who first suggested to the government that the work on the nuclear pile and the nuclear bomb should be conducted in secrecy. This action alone says much about their understanding of the sensitivity of their work.

After the Manhattan Project had concluded its task, its director, Robert Oppenheimer, became the target of many accusations of treason and untrustworthiness, with the result that he was subsequently removed as advisor to the government on nuclear physics. His disgrace took place during the height of that period of paranoia in the United States that had been fostered by Senator Joseph McCarthy.

To fully understand the accusations against Oppenheimer, it helps to know a little about his background. He was born and obtained his first university degree in the United States, but then left to obtain a doctoral degree with Max Born in Göttingen, Germany, in the latter half of the 1920s. There he met many outstanding physicists as well as excellent fellow students, one of whom was Fermi. Afterward he spent time in Zürich at ETH before returning to the United States before the outbreak of World War II. Oppenheimer was chosen to head the Manhattan Project because he was an American-born citizen, while Fermi was not. At that time, the government did not care that Oppenheimer had been educated outside the United States.

After World War II, the American military wanted to continue developing nuclear weapons, including the fusion bomb. Oppenheimer was opposed; such weapons would be even more destructive than the fission bombs that had been used on Japan.

Oppenheimer wished to convince the many governments of the world that these new nuclear weapons were so destructive that the whole world should consider what might happen if they were used in full-scale war. His opposition to the further development of nuclear weapons made him the target of some critics who supported continued research. These critics found out that, in his youth, Oppenheimer had associated with members of the Communist Party; that, and the fact that he had obtained much of his advanced education outside the United States, was used to suggest that he was untrustworthy. Although he had quickly denounced the Communist Party once he had found out its true intentions, he was still suspected of treason and put on trial. As for Oppenheimer's education in Germany, it should not have been construed as un-American; he would have been foolish to pass up such a splendid opportunity to study with Max Born, who was, at that time, one of the very best physicists of quantum mechanics.

Fermi himself called the only press conference of his life to ridicule these accusations against his longtime coworker. Although Oppenheimer was not found guilty of treason, he was dismissed from his post. To his death in 1954, Fermi stood by Oppenheimer and the right to open and free investigation of scientific questions, as well as the right to warn the government and public of the possible dangers of such investigations.

The Direct Detection of the Neutrino

After World War II, scientists, hoping to push back the frontiers of knowledge still further, returned to questions of pure research. In

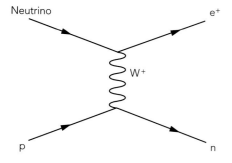

The interaction of a neutrino with a proton.

that spirit, a team led by Clyde L. Cowan and Frederick Reines created Operation Poltergeist, an experiment designed to detect the neutrino directly. Before Cowan and Reines could begin their search, they first needed a source that would provide a large flux of neutrinos—a nuclear-bomb explosion or a large nuclear reactor would suffice. Originally, Cowan and Reines had obtained approval to perform their experiment during a nuclear-bomb test explosion, one of many carried out after the war. However, this method was deemed more difficult than simply using the large new nuclear reactor at the Hanford Research Center in Washington state. Cowan and Reines also needed a clean method of identifying neutrino signals, one that would not confuse real signals with other radioactive decays that normally occur in a reactor.

Conceptually, detecting neutrinos should be simple. Whatever process goes forward in time must have an inverse, which can be described as the identical process going in reverse. The neutrino was expected to interact with matter in an inverse beta-decay reaction: a neutrino would interact with a proton, producing a high-energy electron from the converted neutrino as well as a neutron. This neutron could then be absorbed by an element, such as cadmium, that has a high probability of neutron capture. The cadmium isotope that is formed upon absorption of the neutron is radioactive and would decay, generating an easily detectable particle. A well-known time interval separates the initial production of the electron and the decay of the radioactively unstable cadmium.

Cowan and Reines had set out to look for events with this special signature—two charged particles separated by a fixed amount of time—indicative of a neutrino interaction. This double signal could not easily be emulated by any other radioactive process, and if by accident it were, an easy calculation could prevent any confusion in the results, as explained in the next paragraph. To detect the neutrino's signature, Cowan and Reines used a large tank of liquid scintillator, a chemical compound that emits short light pulses when a charged particle passes through. They lined the tank along its sides, top, and bottom with photomultiplier tubes, devices that are very sensitive to the scintillator's small light pulses. After testing their apparatus with known interactions, they were ready to carry out their experiment near a strong source of neutrinos.

The experiment first ran at the Hanford reactor in 1953. There Cowan and Reines saw the suggestion of a signal, but they needed further confirmation. When they ran an improved experiment in 1956, their results were better: they observed neutrinos at the rate of 0.56 count/hour with a statistical error of 0.06 count/hour. In experimental science, every measurement has a built-in margin of error, usually due to technical limitations (no experiment is perfect) or an accidental simulation of the expected signal. In this experiment, the statistical error is easy to understand. If many single electrons are detected from extraneous decays, some could easily appear to be the double signals if two of them accidentally come close enough in time. A second charged particle might accidentally get through the detector during the window of time after the first signal when the second signal is expected. The "accident" would therefore look like a double signal. Provided that the counting rate of individual particles is known, it is easy to calculate the probability that a second particle, uncorrelated with the first, might arrive with the expected delay time. As long as the number of double signals is higher than the number of such signals predicted to occur by accident, physicists can be confident that their detector has caught neutrinos.

Experimental scientists also like to perform double checks, slightly changing the experiment in some way to see whether they still observe the expected effect. If they do, then they can be more confident that the experiment was performed correctly. To double-check their result, Cowan and Reines ran the reactor at its highest power setting. They expected to see a corresponding increase in the neutrino counting rate since neutrinos are produced from fission fragments that undergo beta decay, and the number of fission fragments being created increases with the power output of the reactor. Indeed, at the highest power, they observed neutrinos at the rate of 2.88 counts/hour with an error of 0.12 count/hour. This signal was 20 times the accidental noise level. For an additional test, they replaced the water in the detector with heavy water, D_2O; heavy water is made from an isotope of hydrogen whose nucleus contains a neutron in addition to the usual single proton. Heavy water slows the neutrons down, making them easier to detect. The result should be twice as many neutrino signals, and, indeed, the neutrino counting rate increased as expected.

Finally, the neutrino had been directly observed. On June 14, 1956, Cowan and Reines sent a telegram to Pauli in Zürich that said, "We are happy to inform you that we have definitely detected neutrinos." After 35 years, Pauli could abandon the self-criticism that he had subjected himself to for suggesting the existence of a particle that he believed could never be experimentally observed.

Even though the Cowan and Reines experiment had double checks, there is nothing like having a result confirmed by another experiment to ensure confidence in the original result. Shortly after Cowan and Reines completed their experiment, others were able to detect the neutrino by alternative means.

While Fermi was teaching in Rome, he had had a brilliant young theory student, Bruno Pontecorvo. Pontecorvo had also left for the

Cowan and Reines in the counting room of their experiment at the Hanford nuclear reactor in Washington state. They were the first to directly detect the elusive neutrino.

Physicist Bruno Pontecorvo in 1955 wearing the Lenin Medal of Honor.

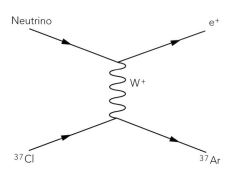

Interaction of a neutrino with a chlorine-37 nucleus, yielding the element argon-37 plus a positron.

United States and had worked on the Manhattan Project with his former advisor. He was mainly a theorist who performed detailed calculations, residing, during most of the war, at the Canadian Reactor Project, which was part of the effort to make bomb-grade plutonium. After the war, Pontecorvo returned to Europe, working first in England and Italy, then eventually making his way to Finland, from where he secretly emigrated to the Soviet Union. Pontecorvo had dreamed of going to the Soviet Union because, as an admirer of the communist system, he thought it less cruel than the capitalist system in its treatment of people. After hearing of Pontecorvo's move to Moscow, Fermi only said that "he carried with him no technical military secrets about the nuclear bomb." Robert Sachs of the University of Chicago disputes this claim because Pontecorvo, during his stay at the reactor where bomb-grade material was manufactured, had become familiar with the xenon problem, a type of contamination that could shut down a reactor. One isotope of radioactive xenon, because of its high cross-section for neutron absorption, accumulating in a reactor core can shut down the fission process if not constantly removed. Solving this problem is a necessary milestone to be overcome in producing large amounts of bomb-grade plutonium. Because of his defection Pontecorvo's ideas were for the most part ignored in the United States, but in Chapters 5 and 7 we will see how some of his ideas were taken up more seriously in Europe by physicists working to achieve a better understanding of the neutrino.

Whatever Pontecorvo's political ideology, he was the first to suggest an alternative technique for detecting the neutrino. Instead of having the neutrino interact with a proton of the hydrogen atom, he proposed having the neutrino interact with a proton or neutron of a heavier nucleus. Then, when a neutrino interaction changes a neutron into a proton, the nucleus of that atom would be changed into another element entirely. Pontecorvo proposed that a large amount of a pure liquid isotope be chemically analyzed over a period of time, all the while watching for an increase in the amount of neutrino-generated new elements.

Ray Davis, a radiochemist at Brookhaven National Laboratory on Long Island in New York, performed just such an experiment based on Pontecorvo's suggestion. Davis looked for neutrino interactions in a liquid sample composed mostly of chlorine. He did indeed observe that the chlorine was slowly being contaminated by argon isotopes created from the chlorine atoms through neutrino interactions. Although the Davis experiment was an excellent confirmation of neutrino detection, it wasn't quite as satisfying as the Cowan and Reines experiment, which was able to see the neutrino interaction as it occurred.

Energy Production in Stars

Our Sun is a large nuclear furnace that, through the process of nuclear fusion, turns elements of small mass—such as hydrogen or helium—into elements of larger mass. Over an 11-year cycle the Sun seems to fade and then burst forth with a flurry of activity. This cycle of change was first detected about seventy-five years ago by botanists who noticed that every eleventh tree ring was very thin, indicating very slow growth, but that in between these low-growth years were years of high growth. In more modern times, these highly active years have been correlated with an increase in sunspots and solar flare activity. Not only do solar flares, emitted from the surface of the sun, cause widespread disruption of radio and satellite communication on Earth, but they damage the protective layers of the upper atmosphere, and can have direct affects on our weather. The actual cause of this cycle is still a mystery, but there is some slight indication that the Sun's neutrino production may follow a similar cycle.

Nuclear reactions in the core of the Sun produce the visible energy we receive from its surface. The immense gravitational pressure in the Sun's core heats up nuclei of hydrogen and helium, giving them

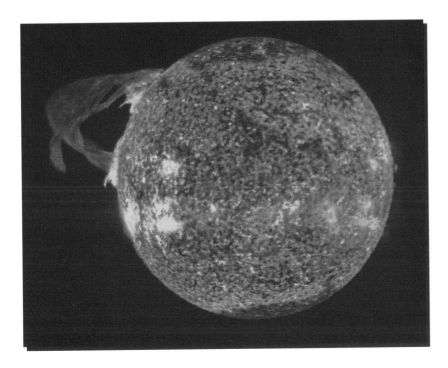

A huge prominence, photographed in ultraviolet light, flares outward from the solar surface. The fusion reactions taking place in the Sun's core are a great source of neutrinos. Each fusion reaction imparts kinetic energy to fusion products and photons; over the course of a million years this energy makes its way to the surface and provides life-sustaining light to Earth.

enough kinetic energy to fuse together. As the nuclei fuse, they release energy in the form of heat and also produce X-rays, gamma rays, and neutrinos.

It is theorized that a star forms out of a large cloud of gas and dust. Slowly, this cloud collapses through the force of gravity into a dense ball. As the gas becomes denser, gravity provides a large pressure in the center of the forming star, causing a high concentration of thermal energy. The energy induces nuclear fusion, which ignites the ball of condensed gas into a star. If a star is formed from an interstellar dust cloud containing heavy elements, Earth-like planets may form in orbit around the star. If the cloud contains only light elements, Earth-like planets cannot be formed.

So a star's life begins with a cloud of gas and dust. If we start our star's life with a cloud that contains only hydrogen and a small amount of helium, several different fusion processes will occur in its core as gravity begins to condense this cloud into a ball. The primary particle interactions at the center of such a young star are:

$$p + p \rightarrow {}^2H + e^+ + \text{neutrino}$$

$$p + p + e^- \rightarrow {}^2H + \text{neutrino}$$

$$p + {}^2H \rightarrow {}^3He + \gamma$$

$$^3He + {}^3He \rightarrow {}^4He + p + p$$

$$^3He + {}^4He \rightarrow {}^7Be + \gamma$$

Here γ stands for a high-energy photon (gamma ray), and Be is the symbol for the element beryllium. The first four reactions are collectively called the pep-cycle. Notice that the lighter elements are burned by nuclear fusion into heavier elements.

Fusion's release of energy guarantees that all the resulting particles carry away more kinetic energy of motion than the initial sum carried by the reacting particles, while momentum conservation ensures that most of this released energy goes to the photon and neutrino. These photons of especially high energy, called gamma rays or X-rays, cannot make it out of the star's interior. Instead, they interact with matter in the outer layers, producing more heat, and that heat is transmitted to the surface, from where it is radiated away as light. The thermal energy released along with the emitted gamma and X-rays makes the Sun glow with the yellow color that we see on Earth. The weakly interacting neutrinos produced in these processes can escape the star completely without interaction.

As a star ages, the heavier elements become concentrated in its core, where they continue to undergo fusion into still heavier elements such as lithium (Li), beryllium, and boron (B) through the following processes:

$$^{7}\text{Be} + e^{-} \rightarrow {}^{7}\text{Li} + \text{neutrino}$$

$$^{7}\text{Li} + p \rightarrow {}^{4}\text{He} + {}^{4}\text{He}$$

$$^{7}\text{Be} + p \rightarrow {}^{8}\text{B} + \gamma$$

$$^{8}\text{B} \rightarrow {}^{8}\text{Be*} + e^{+} + \text{neutrino}$$

$$^{8}\text{Be*} \rightarrow {}^{4}\text{He} + {}^{4}\text{He}$$

These processes also lend kinetic energy to the decay products and the emission of gamma rays; the energetic particles and gamma rays in turn both heat the star and eventually provide a source of energy for the light emitted from the star's surface. The neutrinos produced in these processes have higher energies than those produced in the fusion of lighter elements. Most of the neutrinos produced in these fusion reactions have much greater energy than even the gamma rays.

Stars like our Sun formed from gas clouds rich in carbon (C) and nitrogen. Other stars are born out of hydrogen and helium only, but create carbon through fusion as they age. In either case, carbon becomes the principal element that accumulates in the core. It, too, burns through fusion, by processes that create nitrogen (N) and oxygen (O):

$$^{8}\text{Be*} + {}^{4}\text{He} \rightarrow {}^{12}\text{C} + \gamma + \gamma$$

$$^{12}\text{C} + p \rightarrow {}^{13}\text{N} + \gamma$$

$$^{13}\text{N} \rightarrow {}^{13}\text{C} + e^{+} + \text{neutrino}$$

$$^{13}\text{C} + p \rightarrow {}^{14}\text{N} + \gamma$$

$$^{14}\text{N} + p \rightarrow {}^{15}\text{O} + \gamma$$

$$^{15}\text{O} \rightarrow {}^{15}\text{N} + e^{+} + \text{neutrino}$$

As a star gets older, the fused nuclei of heavier elements accumulate in the center. Shells containing different elements created in the major fusion-burning cycles develop around this core in an onionlike structure, with the heaviest elements in the center. This process continues until the

Theoretical solar models predict the number and energy of neutrinos produced by different fusion processes in the core of the Sun. Physicists can check these predictions by counting the numbers of neutrinos observed in special detectors consisting of elements that are highly reactive with these particles. The bars at the top of the graph indicate the sensitivity of different elements to neutrinos of different energies.

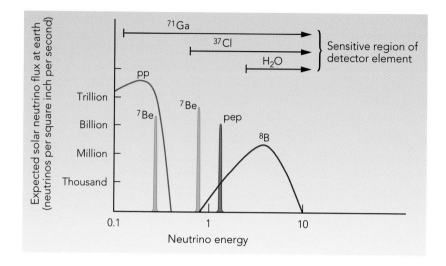

core of the star becomes so unstable or so heavy and dense that it collapses. The eventual fates of different kinds of stars will be discussed in Chapter 8. For now, the important thing to note is that as a star runs out of one principal element, it begins to burn those elements that have accumulated in the core, and in burning those elements, it inevitably releases neutrinos. Since the neutrinos are only weakly interacting, they pass unhindered through large amounts of matter. If we could detect some small fraction of our Sun's neutrinos on Earth, we would, in effect, be observing the nuclear processes occurring in the core of the Sun. Detecting solar neutrinos would be much more valuable to our understanding of the Sun than simply observing the light output from the surface, which does not contain as much information about the solar furnace and its intricate mechanism.

Before leaving this topic a word of praise must be given to the astrophysicists who formulated the solar model, too numerous to completely list. Through the efforts of early theorists like Hans Bethe, who started thinking about the subject in 1939, a precise model of the fusion reactions that are occurring in the Sun's core began to take shape. Their work was complemented by nuclear accelerator experiments, such as those done in 1957 at the California Institute of Technology with William A. Fowler as well as many others afterward, designed to study the likelihood of these interactions occurring at various energies. As a result of these efforts, both experimental and theoretical, today we know that the ^{3}He + ^{3}He reactions account for more than 80% of the Sun's energy.

Research in this field continues still. For example, the role played by very short-lived radioactive nuclei is practically unknown, even though short-lived nuclei such as $^8Be^*$ must contribute in these fusion cycles to produce high-mass elements such as the carbon, oxygen, and nitrogen that we observe around us. So there is still the possibility that some new-found interaction will change our perspective on the fusion reactions in the solar model.

The Case of the Missing Solar Neutrinos

Astronomers generally believe that our Sun is a young to middle-aged star, burning mostly hydrogen and the slightly heavier elements helium, beryllium, and boron. By using models for a young star, it is possible to

The first solar neutrino observatory, located in the Homestake gold mine of South Dakota, was a large tank filled with chlorine. The number of neutrinos found to have reacted with the chlorine was four times less than the number predicted by solar models.

The Homestake solar neutrino detector with its large tank of 100,000 gallons of cleaning fluid (perchloroethylene) rich in chlorine-37. The other equipment is for chemically analyzing the liquid for the presence of argon-37 atoms formed by neutrino interactions. Helium gas is flushed through the fluid to remove the argon atoms so that they may be counted.

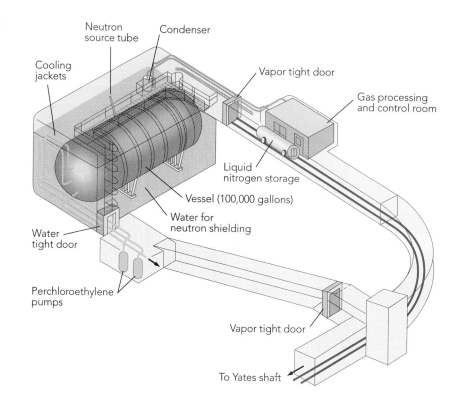

predict the distribution and abundance of neutrinos of differing energies coming from the Sun, based on the different reaction rates we expect to be occurring in the core. The neutrinos of highest energy, which are the easiest to detect, should arise from the decay of boron-8.

Despite many difficulties, Ray Davis, the radiochemist who confirmed the direct detection of the neutrino, has been trying to measure neutrino production in the core of the Sun. He has been hoping to catch neutrinos emitted at the end of the chain of reactions that produces boron-8, when that element subsequently decays. As these neutrinos pass through a large tank containing 100,000 gallons of cleaning fluid rich in the chlorine-37 isotope, which has a high cross-section for intercepting neutrinos, some should be stopped when they interact with the chlorine-37 isotope, converting the isotope into argon-37. Ray Davis and his colleagues designed their experiment to watch for the appearance of this specific isotope of argon in the chlorine solution. To reduce interference from other sources of charged particles, which occur naturally on the surface of the Earth and may cause chemical changes to the chlorine, they placed the tank deep underground in the Homestake gold mine of South Dakota. By circulating the chlorine through chemical filters, which collect the nonchlorine contaminates, and then counting the exact number of argon-37 atoms found per day, they expected to observe the rate of neutrinos coming directly from the core of the Sun.

The Sun is a more immense source of neutrinos than even the most powerful human-built reactor, but it is much farther away and the neutrinos from the fusion of light elements have lower energy than neutrinos produced in reactors. For these reasons, the chance of converting a solar neutrino here on Earth is extremely small. After correcting for the size of their detector and the known properties of neutrino interactions measured in their experiment, the Davis group expected to see one solar neutrino per day. Much to their surprise, they saw only one solar neutrino every fourth day, one-fourth the expected value. There were three possible explanations for the discrepancy: the experiment was incorrectly counting the neutrino rate, the neutrino has some still-hidden mysteries, or our Sun does not function as the solar model predicts. Since the experiment first took data in 1968, it has been refined in an attempt to improve the results, yet the problem of the missing solar neutrinos still remains.

Other experiments were needed to confirm these startling results, and scientists became interested in building detectors that employed different elements to catch the neutrinos. A Japanese experiment, called Kamiokande, used a tank filled with water, which is sensitive only to neutrinos with the highest energy coming from the boron-8 chain. Their limited results concurred with the Davis results achieved using chlorine. With only a five-year sample of data they were able to see a possible correlation between the observed neutrino rate and the solar cycle of sunspot activity.

The latest round of experiments, begun in the 1990s, are designed to look for lower-energy neutrinos. They include SAGE, a Soviet-American collaboration, and GALLEX, a mostly European collaboration that includes Brookhaven National Laboratory. Both are using a rare element, ^{71}Ga (gallium), which is sensitive to the lower-energy neutrinos expected from the fusion of two protons. Neutrino interactions change gallium into germanium in a process similar to the transformation of chlorine into argon. The germanium decays, and its decay products can be detected in the gallium sample. Preliminary results show that the number of solar neutrinos is still deficient by about a factor of two, even though these experiments are sensitive to different neutrinos than the chlorine and water experiments. They tell us that the concerns first raised by the Davis solar neutrino experiments have not been resolved. Thus there may be a slight flaw in our understanding of solar energy production.

The filtration system used in the GALLEX experiment. Germanium is extracted in these chemical analysis columns once every month so that physicists may count the germanium atoms formed by neutrino interactions.

Given the high degree of certainty in the results from the latest experiments, it is reasonable to examine our solar models and the expected fusion-reaction processes for any errors. Historically, physicists and astrophysicists have explored many possible ways of modifying the solar model to account for this discrepancy. I shall cover several of the most discussed options, but keep in mind that it is not certain yet if any of these effects are the actual culprit or if the answer lies elsewhere in another branch of physics.

A good process to begin with is the reaction of boron-8 in the core of the Sun. Fusion reactions were suspected of being strongly dependent on temperature: they could be occurring much less frequently than expected if the Sun's core temperature is slightly lower than models predict. Yet if that were so, we would expect to see a slight decrease in the total light output from the Sun. However, even though light from the Sun is the only other physical quantity we can measure that is also dependent upon nuclear reactions in the core, most of the light we see today is coming from fusion processes that occurred millions of years ago. In contrast, neutrinos coming directly from the Sun's core are an indication of conditions there right now. It is therefore difficult to correlate the amount of light produced by the Sun with the number of neutrinos produced, and thus difficult to double-check the neutrino experiment results and, likewise, the solar models.

In an effort to better measure *current* light output from the Sun, NASA has recently launched a probe called Ulysses that will orbit the Sun to study its overall light output. Because most of the light falling upon the Earth comes from the Sun's equator, Ulysses will travel primarily over the Sun's poles. This mission may in five to ten years provide more data with which to make a better estimate of the Sun's actual energy output.

The gallium experiments, which are particularly sensitive to neutrinos from the fusion of two protons, are less dependent on solar-core temperatures, and agree slightly better with the solar models. However, even these new measurements do not agree with the models sufficiently well and have not produced a complete understanding of our Sun's energy production.

Most solar models assume that our Sun formed from a cloud that was mostly hydrogen; however, we know that cannot be completely correct or there would have been no material from which to create the planets in orbit around the star. The amount of heavy element contamination in the central core of the Sun could, possibly, have an effect on these models. Scientists tried to modify the treatment of these elements in their solar model so that it would better agree with these solar neutrino measurements, but they found little effect from these contaminations. Although heavy element contamination does not offer an explanation of the missing solar neutrinos at present, it is still an option open for further consideration.

In searching for an answer to the question of the missing solar neutrinos, nuclear physicists also considered the role of the many short-lived radioactive isotopes produced in the Sun's core by fusion.

Normally, short-lived nuclei are not factored into calculations, and their cross-section for fusion interactions is not well known. In the Sun's core, where gravitational pressure and densities are high, it is quite possible that these unstable nuclei, too, may be playing an important role, just as in a nuclear reactor one isotope of radioactive xenon can shut the operation down if not removed. Today at Riken Laboratory in Japan, a special accelerator that handles radioactive ions is just coming into operation; it will measure the cross-sections of many short-lived radioactive nuclei that may be found in the Sun's core to confirm whether such nuclei can have an effect on fusion processes.

Scientists have even considered the question of whether our Sun is dying inside, as might be suggested by a lack of neutrinos. Obviously the workings of our Sun are not just a curiosity for scientific investigation—they can also provide useful knowledge for predicting the future of life on Earth. Happily, today we believe the Sun to be in vibrant good health. The solar neutrino discrepancy may be more related to a misunderstanding of nuclear and particle physics. Particle physicists are looking into how neutrino interactions with large amounts of matter may produce other effects that could influence the neutrino sample coming from the Sun, but we must wait until the end of Chapter 6 to learn of this work. Despite the various interesting solutions that have been proposed, the missing neutrinos are a mystery yet to be solved by science.

· · ·

The Manhattan Project's nuclear reactor provided scientists with a source of neutrinos that would eventually permit them to detect this elusive particle. This great scientific project also built the atomic bomb that brought a prompt termination to the war with Japan. Many of the physicists who had helped develop the new technologies during the war had the good sense to try to reduce the threat of nuclear war and work for peace afterward. Their efforts and the openness with which scientists conduct their research raised doubts about the loyalty of many of the scientists, doubts that are occasionally still expressed. In today's experiments, however, scientists from America, the former Soviet Union, and Europe are working together in a new spirit of cooperation.

The detection of the neutrino in 1953 and its confirmation by 1956 showed that Pauli's mythical and undetectable particle was, in fact, real and observable. Although Enrico Fermi did not live to see the neutrino detected, Wolfgang Pauli was still alive and happily received the news. The newly developed techniques of neutrino detection turned out to be useful in experiments to detect neutrinos from the Sun, through which astronomers and physicists hoped to confirm the theorized fusion reaction rates in that star's core. They discovered that the Sun does not produce the exact number of neutrinos they had predicted, and are still working to understand this discrepancy in the aim of achieving a correct scientific model.

A neutrino interaction produces hundreds of different particles. A zoo of exotic particles, including lambda, cascade, sigma, pion, and kaon particles—along with more commonly observed electrons and photons—all leave their tracks in this image, taken in a bubble chamber at CERN.

CHAPTER 5

Additional Neutrinos and Strange Matter, Origins Unknown

By the late 1940s, it seemed that scientists had only one last particle to find—one last piece of the puzzle—and their understanding of elementary particle physics would be complete. A candidate for this last particle was discovered, but shortly afterward the unthinkable occurred—dozens of other new and unexpected particles were also observed. It would take a new approach to particle physics to bring order and understanding to this unwieldy "zoo of many particles."

At this point in our story, physicists knew that matter appears as protons, neutrons, electrons, and neutrinos, only four particles in all. They also knew how these particles are distributed in atoms: protons and neutrons reside in varying quantities in the nuclei; electrons are bound to these nuclei by the electromagnetic force. By the late 1930s, physicists believed that all they had left to discover was the nuclear-force propagator particle that kept an atom's protons and neutrons tightly bound together and the newly proposed weak-force propagator particle. Yukawa thought he had the answer. He predicted the existence of a single particle that would not only mediate beta decay—by coupling to electrons and neutrinos—but would also bind protons and neutrons in nuclei. Discovering Yukawa's particle was the goal that particle physicists set out to achieve.

New Particles from the Cosmos

The Earth is constantly bombarded with cosmic rays, which are simply high-energy particles from deep space. Since their discovery in 1912, scientists had been studying these particles extensively in order to understand their origins. The high energies of cosmic rays also made them valuable in particle physics experiments, for even the best particle accelerators of that time and even today could not produce particles with energies to match those of cosmic rays. In 1937, cosmic ray physicists J. C. Street and E. C. Stevenson found a new particle among the cosmic rays they had observed. This particle had a mass in between that of the electron and proton; moreover, it sometimes appeared with a positive charge state and sometimes with a negative one. It was suggested that this new particle could be Yukawa's particle, since his particle was also expected to be both positive and negative. If an experiment could prove this hypothesis correct, then physicists would have a complete picture of particle physics and the constituents of matter.

To detect cosmic ray particles, Street and Stevenson used an experimental apparatus called a cloud chamber. This instrument is a hollow vessel filled with a gas and fitted with a piston. Pulling the piston out rapidly lets the gas in the vessel expand. Any charged particle that happens to be passing through the expanding gas will create a glowing path of visible ionization. This type of particle detector worked well, but at first the scientists who used it had no way of knowing when a particle was traveling through the gas. Whether they moved the piston at the right moment was a matter of luck. Street and Stevenson needed some way of knowing when to trigger the cloud chamber, and they found it in the use of the proportional counter—a kind of Geiger counter that also recorded pulse height. They placed two proportional counters above the cloud chamber and two below, with the counters below being separated

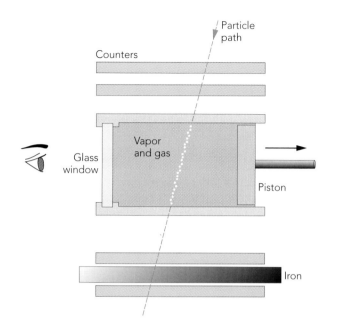

Counters

Particle path

Glass window

Vapor and gas

Piston

Iron

A triggered cloud chamber experiment. When the electronic counters above and below the chamber indicate that a charged particle has passed through, the experimenters pull out the piston of the cloud chamber, causing the gaseous vapor to glow along the path of that particle and any others present. Although the particle is long gone by the time the piston is removed, the trail of ionized particles remains.

from each other by a large amount of iron or lead, penetrable only by a very high energy particle. The four counters, when set off together, acted as a trigger that set off a signal only when cosmic rays of high energy were passing through the detector. Although the particle would have exited the cloud chamber by the time the trigger was activated, its trail of ionization would remain.

Street and Stevenson placed the cloud chamber in a magnetic field so that they could observe the curvature of a particle deflected by the field—a way to identify the mass of a passing particle. Another method of identifying such a particle's mass was to trace the amount of ionization in the cloud chamber along the path of the particle: a heavier particle leaves more ionization along its trail, so the track appears both brighter and thicker. After studying many pictures of cloud chamber events, Street and Stevenson were able to identify the ionization trails of protons and electrons, and eventually they identified a trail that seemed to fall between the two in amount of ionization. This trail was the signature of the new particle, which they named the pion. The particle's direction of curvature in a magnetic field revealed that it could be either positively or negatively charged.

If this new particle were truly Yukawa's proposed force mediator, its decay should look like a beta decay. The energy available for a beta decay is not sufficient to produce a real W particle, and so the particle is virtual in all known beta decays and immediately decays to create an

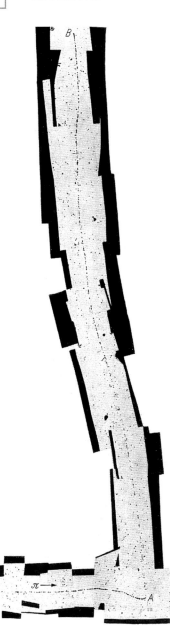

The decay of a pion captured on emulsion film. The pion (π) enters from the bottom left; at the kink (A), it decays into a muon, whose path extends upward. An unobserved neutrino is emitted opposite to the pion.

electron and neutrino pair. If the W particle were to be produced with sufficient energy, however, it would propagate through space for a short time as a free particle, eventually decaying at a greater distance from its source. As usual, it would decay into an electron and a neutrino, but this time the decay products would be of higher energy, obtained from the W's real mass.

Carl D. Anderson and S. H. Neddermeyer observed the decay of the newly discovered particle in a stack of photographic film that had been left exposed to cosmic rays for months atop a high mountain range. Leaving a stack of film where it could record any passing high-energy particles had long been a favored technique of cosmic ray physicists in the early days of particle physics. Physicists chose mountain tops as the optimal site for these experiments—not because they wanted to develop their mountain-climbing skills, but because the Earth's protective atmosphere prevents most high-energy particles from reaching its surface. As a consequence, these particles are more abundant at higher altitudes.

After retrieving the stack of film, Anderson and Neddermeyer had it developed back in the lab. The film revealed faint traces where the charged particles had passed through. By analyzing these tracks in the emulsion stack, the two physicists could study many particle interactions and decays. They identified particles with different masses by the brightness of the traces left in the film, similar to the way in which Street and Stevenson had identified particles in a cloud chamber. But because a stack of film has a higher density than gas in a cloud chamber, it can sample a much larger number of particles. Among their plentiful samples Anderson and Neddermeyer identified the newly discovered pion particle and confirmed its mass as lying between that of the electron and the proton. They were fortunate, also, to observe a few pion decays. The decay of the new particle left a single track with a kink. A neutral, unobserved particle must have been emitted at the kink; otherwise, momentum would not have been conserved. They interpreted the kink in the track as the point in the decay at which an unobserved neutrino was emitted.

These decay events appeared to be consistent with the behavior expected of the weak-force mediator particle proposed by Yukawa, yet there was a problem. The charged particle that appeared in the decay after the kink in the track was not the electron. It, too, was a new particle, also heavier than the electron. This new particle was named the muon. Physicists later learned that the muon behaved as if it were an electron, and they described it as a heavy electron in the early literature. The muon found in the decay of the pion was an unanticipated particle, the very existence of which prompted the famous physicist Isidor I. Rabi to yell out during the conference presentation, "Who ordered that?" No theorist had yet predicted any such particle; this heavier version of the electron took Rabi and the physics community completely by surprise.

In 1940, the Japanese physicists S. Tomanago and T. Araki began an experiment designed to settle definitively whether the new pion was the Yukawa weak-force mediator particle. They proposed that, if the pion were the Yukawa particle, a positive pion should stop in matter differently from a negative pion. A negative pion should sometimes be captured by a nucleus after it is absorbed into an atomic shell, replacing one of the atom's electrons. The positive pion, on the other hand, should not be absorbed into an atomic shell since it does not have the same charge as the electron; instead, the positive pion would decay in flight. A negative pion, because it is much heavier than the electron it replaces, would exchange places with the electron nearest the nucleus. There, the orbit of the pion would take it through the nucleus. The scientists proposed that at that point the pion and nucleus would interact. This was a reasonable conjecture: if the pion was the Yukawa particle, then it should interact with nuclear matter through the weak-force interactions.

The scientists performed the experiment with both iron and carbon targets. When they used an iron target, the negative pion behaved as predicted. But when they used a carbon target, the negative particles were not absorbed and, instead, decayed in flight like the positive pions. In 1947, Tomanago and Araki concluded that the pion was not the weak-force mediator particle proposed by Yukawa, but something unexpected, origins unknown.

Two Different Types of Neutrinos

A reasonable question to ask is whether, when the pion decays into a muon and a neutrino, this neutrino is identical to the neutrino produced by normal beta decay. The discovery of the pion both raised the question and provided the means of answering it, for physicists found that they could make a neutrino beam convenient for study from a pure beam of pions if the pions were permitted to decay into neutrinos and muons. The neutrinos would enter a detector in the path of the beam, whereas the charged muons could be deflected from their straight path by a magnetic field or stopped with a shield so that they did not enter the detector and confuse the picture. In 1959, Bruno Pontecorvo was so certain that the pion-decay neutrino was different from the beta-decay neutrino that he initially called this particle the *neutrina*, the feminine form of "neutrino" in Italian. Pontecorvo suggested trying to observe the pion-decay neutrino in a particle detector as a way of confirming his suspicions. The European Center for Nuclear Research (CERN) in Geneva, Switzerland, considered the experiment of such great importance that it made the project its primary mission. An experiment led by Gilberto Bernardini was begun in 1960.

By that time Melvin Schwartz, then a professor at Columbia University, had proposed a similar experiment for Brookhaven National Laboratory in the United States, and a race ensued between the two labs. At both laboratories accelerator machines of similar energy and design were used to create the pion beams. An accelerator consists of two major components, a magnet that bends the trajectory of the charged particles and a device along the path of the particles that accelerates them. After a beam of protons is placed into the vacuum of the accelerator from a simpler accelerator acting as a source, it travels down the circular beam pipe, curving in the magnetic field. At one or more places in the accelerator ring there are special cavities where radio-frequency energy is pumped in, such that each time the protons pass through this accelerating cavity they are given a little higher kick of energy. In modern accelerators such as the Brookhaven machine, which is still a very useful accelerator today, the beam is focused to keep the protons in a tight bunch, and every time a small kick of energy is given to the proton bunch, the magnetic field is ramped up just enough to keep the bunch in the center of the vacuum pipe. With each rotation around the ring the energy continues to increase and the magnetic fields must be ramped to higher strength to keep the beam positioned. The highest energy achievable in any such accelerator depends directly on how high the magnetic field can be pushed. The accelerator energy is important because many particles of unusual properties can be created for study only at high energies; in particular, the more massive a particle, the higher the accelerator energy must be to create it.

The proton beam produces pions when it hits a target. CERN aimed its proton beam at an internal target in the accelerator ring; the protons near the edge of the bunch passed through a thin layer of metal wires, producing pions. The pions then decayed into muons and neutrinos, which are emitted uniformly in all directions. A bubble chamber nearby recorded neutrino interaction events. In many ways a bubble chamber is similar to a cloud chamber; the major difference is that a bubble chamber uses a liquid as the detection medium, while a cloud chamber uses a gas. A bubble chamber also has a larger volume and records a particle's position with much greater accuracy. Brookhaven opted to use their accelerator to produce an external beam of pions; their beam entered a type of detector called a spark chamber. Made mostly of metal, a spark chamber can have a much higher mass than a bubble chamber and consequently can produce more neutrino interactions. Although from the start it was obvious that the Brookhaven experiment was better, the CERN experiment was expected to obtain results more quickly.

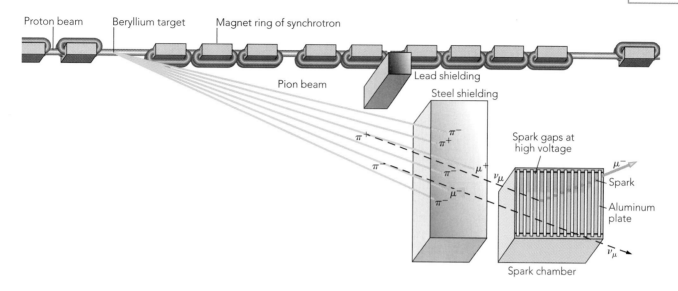

Proton beam | Beryllium target | Magnet ring of synchrotron

Pion beam

Lead shielding

Steel shielding

π^+

π^-
π^+

π^-

μ^+

π^-

ν_μ

μ^-

π^-

Spark gaps at high voltage

μ^-

Spark

Aluminum plate

$\overline{\nu}_\mu$

Spark chamber

The two-neutrino experiment at Brookhaven National Laboratory. The proton beam impinging on a beryllium target creates a beam of pions (π) that decay in flight, forming neutrinos (ν) and charged particles. The charged particles are stopped by shielding, while the neutrinos propagate into the spark chamber, which detects muons or electrons produced by neutrinos in the inverse interaction. When a beam of pure pions was used, only muon (μ) tracks were observed in the spark chamber, indicating that the neutrinos from pion decays are different from the neutrinos produced in normal beta decay.

Although CERN had set a course that should have ensured it would be taking data within a year, an unexpected problem arose. The number of measurable pions from the internal target was insufficient to achieve a workable experiment. This difficulty was not apparent until just before the researchers were ready to begin taking data; however, when it became clear that the accelerator provided drastically fewer neutrinos than expected, the director general of CERN, the American physicist Victor Weisskopf, decided that the experiment should be terminated.

At the same time eight other physicists were preparing to carry out the similar experiment that had been planned for Brookhaven. They included Schwartz, Leon Lederman, and Jack Steinberger, a professor at Columbia University who had also participated in the initial CERN experiment. This experiment would avoid the problems that the CERN experiment had encountered: the physicists planned to create more neutrinos, from an external beam, and more neutrino interactions. Instead of a bubble chamber, they relied on spark chambers—which, like bubble chambers, record single events on film. But to increase the number of interactions, they would place aluminum plates between the individual planes of the spark chambers, in this way providing more target nuclei with which neutrinos could interact. This adjustment, along with their extracted neutrino beam line, would ameliorate the rate problem and make the experiment feasible.

The Brookhaven experiment, using protons from the AGS accelerator at 15 GeV, provided a beam of high-energy pions. The beam of pions was allowed to decay as it continued to travel over a certain length.

All the charged particles were stopped by shielding at the end of the beam line, leaving only a neutrino beam, which continued through the shielding to the other side and into the spark chambers. The neutrino interacted in the aluminum plates between the spark chamber gaps, producing charged particles that were observed on film as they passed through the spark gaps, emitting light. Any charged particle that entered the detector from the outside was excluded from consideration in the analysis. The spark chamber's response to muons and electrons had been previously studied, and it was known that muons would penetrate the many layers of the spark chamber, leaving a single clean track of sparks. Electrons, however, could not penetrate more than a few layers before they would shower into multiple, many-charged particle tracks.

The experimenters had now recorded many neutrino interactions. How would their recordings tell them whether there were two types of neutrinos? Since the experiment had started out with neutrinos from pion decays, which were then converted into observable particles through the inverse interaction, all the experimenters had to do was identify the particles existing after the neutrino interactions. In the case of pion-decay neutrinos, this inverse reaction would produce muons, while in the case of beta-decay neutrinos, the inverse reaction would produce electrons. If there was only one type of neutrino, the neutrinos interacting in the spark chamber should produce both types of inverse decays, sometimes creating muons and sometimes electrons. After recording many neutrino interactions in the spark chamber detector, the experimenters found that only muons were produced. Since no events showed any sign of electrons, the physicists concluded that the neutrino emitted opposite the muon when a pion decayed was distinctly different from the neutrino emitted opposite the electron of beta decay.

It seems, therefore, that the neutrino comes in two varieties. One variety is called the electron-type neutrino because it is emitted in beta decay along with an electron. The other variety is called a muon-type neutrino because it is emitted in pion decay along with a muon. These neutrinos and any particle associated with them are very different from the particles emitted in the beta decay of normal matter. The discovery of the two types of neutrino essentially marked the birth of a new era of particle physics, as physicists came to recognize the existence of new "species" of matter, not present in what most of us think of as normal, everyday matter. The neutrinos were just the first of this "zoo" of particles awaiting discovery.

An interesting story about this experiment was told to me by Steinberger. None of the experiment's three leading physicists were actually working for Brookhaven—rather, they were all at Columbia University. Steinberger could have worked for Brookhaven if he had been willing to sign an oath of loyalty to the United States government swearing that he was not a communist. He had refused to sign such a

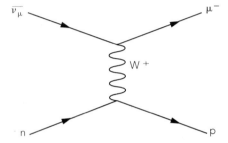

A Feynman diagram of a muon-type neutrino ($\overline{\nu}_\mu$), produced by pion decay, interacting with a neutron from an aluminum nucleus. The interaction produces only a muon, represented by the Greek letter mu (μ). The Greek letter nu (ν) is the symbol for the neutrino; the type of neutrino is indicated by the subscript.

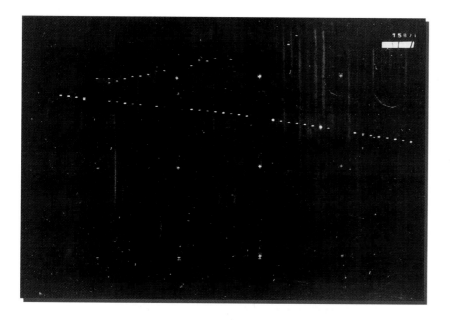

The long bright track in the middle is the trace of a muon produced by a neutrino interaction in a spark chamber. An electron, in contrast, would have showered into many tracks after passing through only a few layers of the material separating the spark gaps. By such differences the physicists of the the two neutrino experiments could tell that the neutrinos from their pion beam produced only muons in their decay and never electrons.

document while at the University of California at Berkeley, where he had been working, because "I was reasonably left (never a communist, even less a party member) and somewhat active in the labor movement during the war years." He and many other physicists declined to sign such documents because they objected to the wording of the oaths and

The eight physicists who performed the first experiment that detected two distinctly different types of neutrinos. From left to right: J. Steinberger, D. Goulianos, J.-M. Gaillard, N. Mistry, G. Danby, W. Hayes, L. Lederman, and M. Schwartz.

because their own open-minded attitudes led them to support the right of others to think as they chose.

During the 1950s and early 1960s, scientists needed a minimum security clearance to hold a job at any national laboratory, or even to perform an experiment there. It was for this reason that Steinberger left Berkeley to take a professorship at Columbia University, a private institution that did not require such an oath. The Atomic Energy Commission put Steinberger on trial and revoked his access to Brookhaven, so he often worked at CERN instead. Eventually the national laboratories dropped the requirement for a security clearance, permitting Steinberger's participation in Brookhaven's experimental program—another example of how the moral standing of a physicist had to be accommodated to achieve important scientific results.

These American physicists, for reasons of their own, did not agree with the views held by the United States government at that time. Since the end of World War II, scientists, along with the rest of society, increasingly questioned their government's true intentions. The attitude expressed in the phrase "my country, right or wrong," which was used to justify cooperation with government or the military, was no longer prevalent. Scientists from around the world began to organize together in groups that spoke with a single voice. They expressed their opinions about the dangers of nuclear war directly to government officials and in publications such as the *Bulletin of Atomic Scientists*, regardless of any potential reprisals by the government. This outspokenness broke with the unwritten rule that scientists should concern themselves only with science and follow, without question, the political party in power in their country.

The actions of American physicists during the postwar years stand in contrast to Enrico Fermi's actions in Italy during the 1920s. In the archives at the University of Chicago are stored the Department of Defense security clearance forms that Fermi filed to work on the Manhattan Project. In them Fermi admits to having been a member of the Italian Fascist Party. He explains that this membership was mandatory. As a professor at an Italian University, he was required to be a member of the ruling political party. Otherwise, he asserted, he would not have joined, and he also stated that he was never an active member. Jack Steinberger, who was Fermi's doctoral student at the University of Chicago from 1946 to 1948, and knew well his mentor's thoughts and feelings, requests that I point out that "Fermi was a really brilliant and organized scientist, who had little energy left over for politics, art, or other culture."

In the new age of freedom in the United States during the 1960s, the attitude of physicists had changed. Physicists such as Steinberger had become more careful and would not simply accept the attitude of the ruling government.

Strange Matter

With the discovery of the unexpected new particles, physicists began to classify them into groups of particles that shared similar characteristics. Particles such as electrons, muons, and the two types of neutrinos are generally called leptons. They have a half-integer spin and obey Fermi-Dirac statistics, which means that two or more leptons may not occupy the same place at the same time. This rule is a direct consequence of the Pauli exclusion principle from basic quantum mechanics, and it means that electrons around a nucleus have to be in different orbits. However, two electrons may occupy the same orbit provided that one has a positive half-integer spin and the other has a negative half-integer spin. Leptons are probably best defined by how they do not interact. They do not interact by the strong nuclear force and so interact directly with protons, neutrons, or pions only by the weak force.

A property of leptons proved especially useful in understanding particle interactions and decays. Lepton number is always conserved: the number of leptons before an interaction or decay must equal the number afterward, and leptons are not created or destroyed. One proviso is that an antilepton counts as negative, so in doing the arithmetic the production of an antilepton cancels out the production of a lepton. Hence, to conserve lepton number, leptons are created in pairs formed of a lepton and an antilepton. The two lepton types demonstrated by the two-neutrino experiment just described—muon leptons and electron leptons—are each conserved separately. In other words, if there is one lepton at the beginning of any interaction or decay, then there must be one lepton of the same type remaining afterward.

Take, for example, the case of a neutron that undergoes beta decay. Because a neutron is not a lepton, the lepton number starts out as zero. The decay produces an electron with a positive lepton number and a neutrino that is actually an electron-type antineutrino with a negative lepton number. The opposite lepton numbers sum to zero—and lepton number is conserved. This type of conservation resembles charge conservation, where positively and negatively charged particles are produced in pairs from the decay or interaction of a neutral particle.

The muon that is produced in the decay of a pion was itself observed to decay into an electron. But since a muon is heavier than an electron, physicists suspected that the difference in mass had been used in the creation of one or more invisible neutrinos. Remarkably, lepton number conservation tells us exactly what the missing particles are. The muon decay must involve the emission of two neutrinos, a muon-type neutrino and an electron-type antineutrino. The electron and the electron-type antineutrino have opposite lepton numbers that add up to zero, so no extra electron lepton number is created. The muon-type

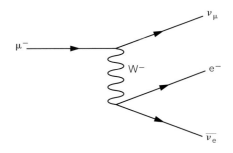

A Feynman diagram of muon decay. Lepton number is conserved both before and after the decay, providing particle physics with a new conservation rule.

neutrino is of the same lepton number as the original muon; the muon lepton number is therefore conserved.

Shortly after the pion was discovered, physicists observed many other new particles with masses distinctly different from that of the proton or electron. These particles were given unusual names such as kaon (K), lambda (Λ), sigma (Σ), and cascade (Ξ). All of these particles were observed to decay into more common types of matter such as the pion π or proton p:

$$K^0 \to \pi^+ + \pi^- \qquad \Sigma^+ \to p + \pi^0$$

$$K^+ \to \mu^+ + \nu_\mu \qquad \Xi^+ \to \pi^+ + \Lambda^0$$

$$\Lambda^0 \to p + \pi^-$$

This proliferation of new particles began in 1947, but it would take several decades to figure out why so many particles existed. The number of new particles was growing so fast that no one believed any longer that they were all fundamental. Explanation and organization was necessary before our understanding of particle physics could advance.

In 1952 Abraham Pais introduced the principle of *associated production* to describe a feature of these new particles. He proposed that the new particles, although seen individually in detectors, are actually produced in pairs. But it was the formalism introduced by Murray Gell-Mann in 1953, when he proposed a new quantum number named *strangeness,* or S, that finally brought some order. The term "strange matter," used to describe the new particles, reflected the sense of confusion and bewilderment that these new forms of matter evoked within the physics community. Gell-Mann's scheme gave zero strangeness to ordinary matter, such as the neutron, proton, or pion, and assigned a strangeness of S = +1 to the positive kaon and a strangeness of S = −1 to the lambda and negative kaon. Gell-Mann then evoked the associated production principle in postulating that new particles must be produced such that strangeness is conserved, just as lepton number is conserved. The sum of strangeness before and after an interaction or decay must be identical. For example:

$$\pi^+ + n \to \Lambda^0 + K^+$$

$$\text{Strangeness: } 0 + 0 = -1 + 1$$

$$0 = 0$$

This theory postulated that the new strange particles were produced by the strong-force interaction. Thus, the weak-force decays of these particles were permitted to violate strangeness conservation. In this sense strangeness conservation is unlike lepton number conservation, and the

1 H 1.00																	2 He 4.00
3 Li 6.94	4 Be 9.01											5 B 10.81	6 C 12.01	7 N 14.00	8 O 15.99	9 F 18.99	10 Ne 20.17
11 Na 22.98	12 Mg 24.30											13 Al 26.98	14 Si 28.08	15 P 30.97	16 S 32.06	17 Cl 35.45	18 Ar 196.96
19 K 39.09	20 Ca 40.07	21 Sc 44.95	22 Ti 47.88	23 V 50.94	24 Cr 51.99	25 Mn 54.93	26 Fe 55.84	27 Co 58.93	28 Ni 58.69	29 Cu 63.54	30 Zn 65.39	31 Ga 69.72	32 Ge 72.61	33 As 74.92	34 Se 78.96	35 Br 79.90	36 Kr 83.80
37 Rb 85.46	38 Sr 87.62	39 Y 88.90	40 Zr 91.22	41 Nb 92.90	42 Mo 95.94	43 Tc 97.90	44 Ru 101.07	45 Rh 102.90	46 Pd 106.42	47 Ag 101.86	48 Cd 112.41	49 In 114.82	50 Sn 118.71	51 Sb 121.75	52 Te 127.60	53 I 126.90	54 Xe 131.29

55 Cs 132.90	56 Ba 137.32	57 La 138.90	58 Ce 140.11	59 Pr 140.90	60 Nd 144.24	61 Pm 144.91	62 Sm 150.63	63 Eu 151.96	64 Gd 157.25	65 Tb 158.92	66 Dy 162.50	67 Ho 164.93	68 Er 167.26	69 Tm 168.93	70 Yb 173.04	71 Lu 174.96	72 Hf 178.49	73 Ta 180.94	74 W 183.85	75 Re 186.20	76 Os 190.20	77 Ir 192.22	78 Po 295.08	79 Au 204.38	80 Hg 200.59	81 Ti 207.20	82 Pb 207.20	83 Bi 208.98	84 Po 208.98	85 At 209.98	86 Rn 222.01
87 Fr 223.01	88 Ra 226.02	89 Ac 227.02	90 Th 232.03	91 Pa 231.03	92 U 238.02	93 Np 237.04	94 Pu 224.04	95 Am 196.96	96 Cm 196.96	97 Bk 247.04	98 Cf 251.07	99 Es 252.08	100 Fm 257.09	101 Md 196.96	102 No 259.10	103 Lr 262.11	104 Rf 261.10	105 Ha 262.11	106 Sg 263.11	107 Ns 262.12	108 Hs 265.13	109 Mt 266.13	110 270.20	111 272	112						

origins of this difference will not be fully explained until the end of Chapter 6.

One hundred years earlier Dmitry Mendeleyev had been able to organize the elements by creating the periodic table; his creation not only helped explain the differences in mass between elements but also helped later physicists see how the elements were constructed from more fundamental particles. Similarly, this new quantum number of strange matter permitted an organization of the new particles that would suggest an underlying structure.

Back in 1932, Heisenberg had suggested a new way of looking at the neutron and proton. He proposed that if the electric charge of these particles is ignored, their nearly identical mass suggests that they are related to each other. This relationship is not through the electric charge, but rather through the strong force that governs nuclear interactions. Heisenberg devised a system according to which the neutron is the opposite partner of the proton; in this system the pair is held together by a strong nuclear force called *isospin*. Just as the positive pion is the opposite partner of the negative pion through the strong nuclear force interaction, the neutron was thought to be the opposite partner of the proton, also through the strong nuclear force interaction.

In postulating this relationship, Heisenberg had introduced a symmetry, to which he gave mathematical form by assigning the proton a $+\frac{1}{2}$ isospin and the neutron a $-\frac{1}{2}$ isospin. The word "isospin," for "spinlike symmetry," actually had nothing to do with spin, but

The periodic table of the elements, which helped scientists develop the theory of protons and neutrons in the atomic nucleus. The number at the top of the box gives the number of protons in the nucleus, while the number at the bottom is the atomic weight, which is an average of the masses of all isotopes found in nature or created in accelerators. The symbol for the element is in the middle of the box; elements 110, 111, and 112 have been identified, but have yet to be named. Elements in the same column have similar chemical properties, but differ in mass.

permitted Heisenberg to use the mathematical symmetry of known spin-ning systems to explain and organize the known similarity of protons and neutrons. Neutrons and protons are certainly not symmetric in all their characteristics, particularly charge, but Heisenberg proposed only that those characteristics governed by the strong force should be sym-metric. The fact that one particle, the proton, is positive in charge and the other, the neutron, is neutral in charge is not a result of the strong nuclear force, but has its origins in the electromagnetic force. This charge difference is also the reason that the neutron and proton do not have quite the same mass.

Gell-Mann's inspiration was to graphically plot the isospin of a par-ticle against its strangeness. By this time not only were particles coming with a strangeness of $S = -1$, but the observed decay of the cascade (Ξ) particle had provided clear evidence for strangeness of $S = -2$. Gell-Mann's graph created a new sense of order. Excluding the muon, the electron, and their associated neutrinos, all the new particles seemed to fall nicely into two categories that could be characterized by the total spin of the particle (the regular spin, not the isospin). One set, called *mesons,* includes the pion and kaon and all other particles with zero or integer spin. The other set, called *baryons,* includes the proton, neutron, lambda, sigma, and cascade particles, which have multiples of half-inte-ger spin. Both these sets are collectively called *hadrons,* and all these particles interact by the strong nuclear force.

Baryons and mesons organized according to isospin versus strangeness.

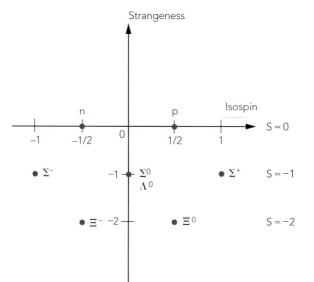

Once physicists had separated these two groups of hadrons and carefully assessed their strangeness quantum number, the graph of strangeness versus isospin made evident a still more fundamental underlying order. Each baryon seemed to be composed of three components, while the mesons seemed to be composed of only two. Gell-Mann named these underlying components *quarks*. Normal matter was proposed to be formed out of only two types of quarks, called up and down quarks. The proton is known today to be the combination of two up quarks and one down quark; the neutron is the combination of two down quarks and one up quark. Strange matter has a quark, called a strange quark, that is similar in behavior to the down quark. The name "hyperon" is used to describe any baryon that is composed of a strange quark.

Originally, physicists used this description of matter as composed of quarks solely for accounting purposes—it was a useful aid in predicting the behavior of matter. However, a debate soon began about whether these particles were actually real. After collecting all the newly observed particles into the organizing graph of strangeness versus isospin, physicists noticed some obvious holes. If the baryons were truly composed of three quarks, then a real particle with three strange quarks must exist. This proposed particle, named the omega (Ω), was to be a test of Gell-Mann's quark composition theory. The omega particle was eventually seen in 1964 in the Brookhaven bubble chamber. Once physicists had actually observed a particle predicted by this new theory, the quark composition model gained credibility. Although some other underlying structural argument might have eventually accounted for the omega particle's nonexistence, the fact that it does exist made life simpler for the theorists who were trying to organize the "zoo" of newly discovered particles.

Constituents of Matter

The complete story behind the discovery of the quark theory of matter, with its many subtle details, is peripheral to the story of the neutrino and therefore beyond the scope of this book. However, because the quark theory is useful for the remainder of the book, I shall provide a summary here, but without trying to justify the logical necessity for its implementation. In this section I will also cover the rudiments of the strong nuclear force and its relationship to the quark theory of matter.

All hadronic matter—particles that are not classified as leptons—is composed of quarks. The quarks are fermions, meaning that they satisfy Fermi-Dirac statistics just as the leptons do. All quarks have half-integer spin. It is the quarks that interact through the strong nuclear force.

The theorist Yoichiro Nambu, Yukawa's student, introduced another quantum number when he proposed that the quarks must possess

A Feynman diagram depicting the decay of a pion into a muon and neutrino at the quark level.

a property called color. Color in this case is not "real" in the sense that a particular kind of quark is actually a particular color. Rather, the concept of color is an analogy used to describe the way in which the quarks group together to form baryons. Baryons, such as the proton or neutron, are composed of three different kinds of quarks, and each quark has a different color. These colors combine to form no color at all, much in the same way that red, blue, and green light when mixed together form white light—the absence of color. And since the theory of strong nuclear interactions requires physically observable particles like baryons to be neutral in color, the color analogy works well with the theory.

Quark color explains the mystery of how protons—like-charged particles that should repel each other—bind together in the nucleus of an atom. Quarks of different color charge are attracted to each other. The three quarks that make up a proton are all likely to be attracted to quarks of different color charge composing other protons. The strong nuclear force of attraction between the quarks is propagated by *gluons*, particles that join quarks together. The three quarks of an individual proton are bound together by the continuous exchange of gluons, which means that the gluons are a substantial component of hadronic matter.

Similarly, whereas a free neutron is an unstable particle that decays, a neutron bound in the nucleus interacts with the protons and other neutrons through the strong force of color charge and remains stable. Mesons, such as the pion or kaon, are also composed of quarks, but in their case each meson is made of a quark and an antiquark. For example, the positive pion is the combination of an up quark with an anti-down quark. Since pions and kaons are physically observable particles, they too must be neutral in color. This means that each member of the quark-antiquark pair in a meson must come in the "anti" color of its partner—for example, a red quark must be paired with an anti-red quark, a combination that is colorless. Again, it should be stressed that this new quantum number of color assigned to quarks has nothing to do with "real" colors, but is used only as an analogy.

Feynman diagrams have been extended to depict interactions governed by the strong nuclear force. Gluons are drawn as spring-type wavy lines that connect two quark lines. Unlike the photon or the W particles, which propagate the electromagnetic and weak nuclear force, the gluons of the strong nuclear force can interact with other gluons. This ability introduces an added complexity, something that we are not going to worry about in this brief description of the strong force. Like the photon of the electromagnetic force, the gluon is required to be massless. The simplest strong nuclear force attraction between nucleons is drawn as the virtual exchange of a gluon between two quarks. For example, when two protons interact by the strong force, shown in the

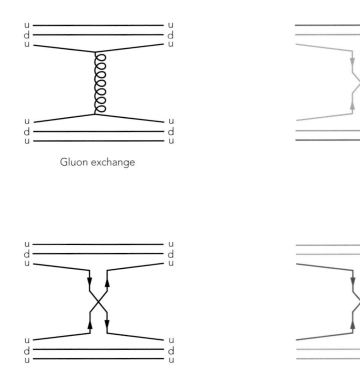

Gluon exchange

Virtual meson exchange

Feynman diagrams for two types of strong interactions between protons, each accompanied by one of many possible color flow diagrams. Notice that the color flow diagrams are of an identical structure.

Feynman diagram on this page, a gluon is exchanged between two up quarks. Each gluon carries two colors; a color-exchange diagram for the interaction can be drawn when the nuclear force is being considered.

The strong force binds two protons together in a nucleus. While quarks in a proton or neutron interact through the exchange of virtual gluons, different protons actually interact through the exchange of virtual mesons. Such a virtual meson is created from the exchange of two quarks in the interacting protons. As the two quarks pass each other, going in opposite directions, they look just like a meson but without the mass that a real meson would have. A careful study of the figure above helps the reader grasp the similarities with gluon exchange; the color flow is identical but not the quark paths. In both cases, the color flows crisscross each other, and because we are drawing them on paper they overlap, but actually the crossed paths in the exchange are not interactions. Although the two processes look identical in the color-flow diagrams, they are actually slightly different since gluon exchange is a short-range force that exerts its influence within an area the size of the proton and meson exchange dominates the long-range interactions that take place over the length of the nucleus.

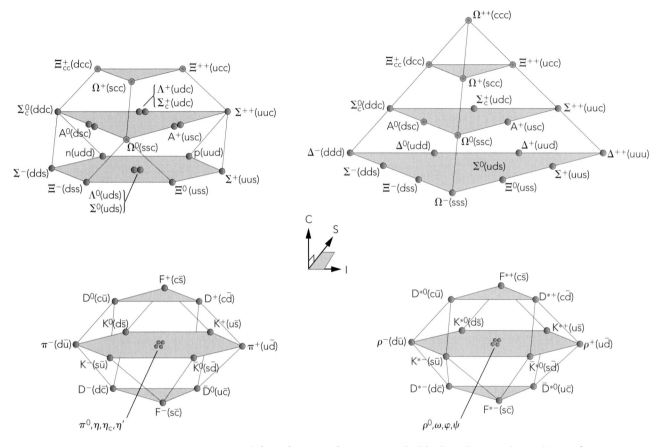

A three-dimensional organizational table shows how quarks combine to form various types of particles. The top two diagrams illustrate the construction of baryons, and the bottom two the construction of mesons. The diagrams on the left are the lowest-mass ground states for the given combinations of quarks. Those on the right are the first excited states of quarks through the strong force; additional graphs would have higher-mass states. Not all combinations of quarks are permitted for the ground-state baryons, while some duplicate quark combinations are formed with quarks of the same flavor but different masses. A subscript c indicates that a particle differs from another represented by the same Greek symbol only in having a charm quark instead of a strange quark. Open circles indicate quark combinations that are expected to exist, but have not yet been observed.

Today we know that quarks and leptons come in many "flavors" such as up, down, strange, and charm; these particles are grouped according to flavor in several generations. The first generation includes the particles that make up all normal matter. The first quark generation consists of the down quark and the up quark, while the first lepton generation consists of the electron and the electron-type neutrino. The

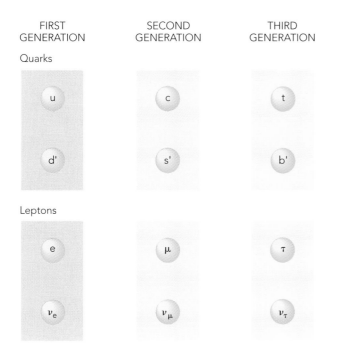

FIRST GENERATION	SECOND GENERATION	THIRD GENERATION
Quarks		
u	c	t
d'	s'	b'
Leptons		
e	μ	τ
ν_e	ν_μ	ν_τ

A table of the constituents—or building blocks—of matter, organized by how the particles pair in weak-force interactions. The prime symbol that appears with the down, strange, and bottom quarks (d′, s′, and b′) will be explained in a later chapter.

electron has a -1 electrical charge and the neutrino has no charge; the up quark has $+2/3$ electrical charge and the down quark has $-1/3$ electrical charge. The discovery of the charm quark, in 1974, completed the second generation, which follows the pattern of the first in including two quarks and two leptons. The quarks are the strange quark and the charm quark, while the leptons are the muon and the muon-type neutrino that were introduced earlier in this chapter. Today, a third generation is also known to exist. Just how many generations of quarks and leptons may exist has been heavily debated, and this controversy will be treated more fully in Chapter 9. Until then we will concentrate on the first two generations of quarks and leptons.

Each new quark discovered, other than the two in the first generation, has been assigned its own quantum number. The strong nuclear force requires that these numbers be conserved, but they may be violated in weak decays. The up and down quarks do not have their own quantum numbers but instead have been assigned $+1/2$ isospin for the up quark and $-1/2$ isospin for the down quark. The assignment of isospin rather than quantum number is unique to the first generation of quarks. No such special feature is required for the leptons or any other quark generation.

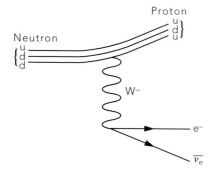

A Feynman diagram depicting the beta decay of the neutron at the quark level.

The weak-force propagator particle W couples the quarks and leptons of any generation. It is through this coupling that we get a glimpse into the origin of the beta-decay mechanism. Imagine a down quark colliding with an anti-up quark; that union would form a virtual W particle that travels forward on the time axis of a Feynman diagram. The W particle then decays into an electron and an electron-type antineutrino, a same-generation pair. The W may couple to a lepton or quark doublet of any generation, in this example giving an electron and antineutrino pair. In terms of the weak nuclear force, electrons are the opposite of the electron-type neutrinos; the two particles are said to be carrying opposite but equal weak charge. This description of the neutron decay tells us what is happening at the quark level, which is the level responsible for the beta-decay process. So, although it is still correct to say that the neutron decays into the proton, electron, and antineutrino, at the quark level we know what is really going on: one of the neutron's down quarks decays into an up quark and a pair of leptons consisting of an electron and an antineutrino.

The doublets of quarks and leptons in each generation demonstrate how these particles are related to each other through the weak force: each doublet consists of particles of opposite weak charge. As we saw back in Chapter 3, only left-handed weak interactions are allowed.

Quark-level Feynman diagrams are the most important kind for physicists. It is only at the quark level that these graphs become useful for calculating weak-interaction physics, since the rules for obtaining proper equations from Feynman diagrams are formulated only for quark-level processes. The equations permit physicists to determine measurable quantities that can be verified in experiments. The same is true for Feynman diagrams depicting strong-force interactions: to be useful for any calculation, the diagrams are always drawn at the quark level.

Originally, physicists postulated that hadronic matter (meaning protons and neutrons) is composed of quarks in order to create a sort of accounting ledger that would help determine which interactions and decays were possible. Quarks began to seem real when physicists observed the omega particle predicted by the theory, but they knew it would be even better to observe quarks themselves. Since quarks have color charge but only colorless particles are observable, observing a free quark in order to prove its existence is impossible. However, scattering experiments performed at the Stanford Linear Accelerator Center (SLAC) in California have allowed physicists to probe the structure of the proton, providing direct evidence that quarks exist. Just as Rutherford used positive alpha particles in scattering experiments to deduce the charge distribution in the atom, a SLAC experiment led by Jerome I. Friedman, Henry W. Kendall, and Richard E. Taylor used very high energy electrons to probe the internal structures of protons and neutrons. By observing the scattering of electrons off protons, the SLAC team was able to deduce that the proton has three charged centers within it, evidence for the true existence of quarks.

. . .

When Street and Stevenson discovered the pion, it appeared to be the last missing particle needed to complete the picture of particle physics. But then scientists saw the pion decay into a heavy type of electron called the muon, along with an unobserved neutrino. It became clear that the neutrino of pion decay was different from the neutrino of beta decay—neutrinos, it seemed, came in more than one variety. The discovery of the muon, the second type of neutrino, and many other unexpected particles collectively called strange matter was a great shock to the physics community. The new particles were all related to each other, but it would take a new way of organizing these particles before that could be understood. This was a golden age of particle physics, when it seemed that every experiment found something new. The most difficult supposition to prove *explicitly*, however, was that the pion-decay neutrino was not the same as the neutrino of beta decay.

Physicists learned that not only do particles carry electric charge, they also have weak charge and color charge. Another property, isospin, is something like spin in the strong nuclear force. Other quantum numbers—such as quark flavor—and the arrangement of quarks into generations were proposed as a means to organize the "zoo" of new particles that had been found. At first the actual existence of quarks was doubted even by the physicists who invented them, but in time experimental evidence confirmed their existence in the subatomic world.

A particle in the subatomic world must observe certain rules. A particle must have integer multiples of charge; no fractionally charged particles can be free. Every real, observable particle must be colorless—either it must be a baryon, which has three quarks of different colors that merge to make a neutral color, or it must be a meson, which has a color and anticolor combination. Particle decays must conserve energy, momentum, and lepton number as well as strangeness and charm. Weak decays, however, not only violate parity, as was discussed in Chapter 3, but other rules as well. Why, is a subtle mystery of weak decay whose solution has yet to be revealed.

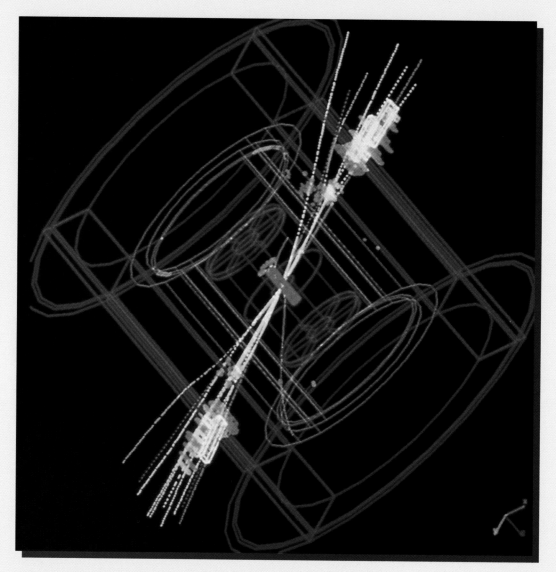

A matter-antimatter collision recorded by the DELPHI detector at CERN.
Matter and antimatter collide at the center of the image and annihilate,
producing a spray of particles away from the collision point, including invisible
neutrinos and antineutrinos.

CHAPTER 6

Can the Neutrino Be Its Own Antiparticle?

Many science fiction writers have called upon antimatter to serve as an energy source for their imaginary adventures in space exploration. Its potential power stems from its nature as a form of matter that is in some sense the inverse of ordinary matter, for when matter and antimatter meet they could be said to "cancel out" each other. Both are destroyed, and their mass is converted into pure energy. The story of how scientists became aware of the existence of antimatter makes an interesting point about the occasional blindness of the scientific observer. Today, scientists are trying to find out whether antimatter can spontaneously change into normal matter. At this point in the story of the neutrino, they are also trying to find out whether the neutrino itself might be the only elementary particle in nature that is its own antiparticle.

Matter and Antimatter

The first hint that such a thing as antimatter might exist came in the early 1930s, when the Cambridge University physicist Paul Adrien Maurice Dirac formulated a relativistic theory of quantum electrodynamics. His was the first theory to use Einstein's new relativity in a quantum theory of elementary particle interactions. The equations Dirac derived were symmetric with respect to the charge of a particle and predicted that a positive particle identical to the negative electron should exist. In the first German edition of Wolfgang Pauli's book on quantum mechanics, Pauli pointed out that if Dirac's equation represented reality, then positive electrons should have already been detected. Pauli's criticism implied that, until such a particle was actually observed, the new relativistic theory of quantum electrodynamics should be viewed with some skepticism. It wasn't long before nature proved both Dirac and Pauli correct.

In 1933, Carl D. Anderson at the California Institute of Technology announced the discovery of the positive electron, now called the positron. After examining more than a thousand photographs of cosmic ray tracks produced by his cloud chamber experiment, Anderson was able to identify 15 tracks whose degree of ionization indicated that the particle had the mass of an electron, but whose curvature in a magnetic field was that of a particle of opposite charge. Many similar cloud chambers must have taken hundreds of "pictures" of this new type of particle, but experimental physicists had paid little or no attention to them. Some physicists had actually noticed the unusual tracks but had explained them away as charged particles going upward through the Earth or coming directly toward the lens of the camera in such a way that their positive-charge curvature was an optical illusion. If only these scientists had stopped to think about the number of unusual events they were recording, they would have realized that the rate of these events could not have been so large.

After the discovery of the positron, Dirac is quoted as having said, "It is very easy only to see what one is told to look for." Experimenters sometimes become so caught up in their study of one particular aspect of physics that new particles or phenomena slip by them because they are looking for something else. Experimental physicists had been using cloud chambers for many years by the time Anderson announced the

The blackboard kept during the Berkeley antiproton experiment. Most of the writing on the board was put there to keep track of the possible antiproton yields for different settings of the experiment. However, a baseball enthusiast should be able to identify the date and time this photo was taken because in the top right corner someone has jotted down the score in the sixth game, third inning, of the World Series between the Yankees and the Dodgers.

positron's discovery. They had simply missed the tracks of the positron because they hadn't been looking for them.

Pauli had strongly objected to Dirac's theories of relativistic quantum mechanics, believing that if Dirac's equations were correct, scientists should have detected the positron long ago. Anderson's discovery of the positron confirmed the theories and assured Pauli of their accuracy. Pauli removed his objections from the second German edition and first English edition of his textbook on quantum mechanics.

Even though the positron's existence was now certain, whether other types of antimatter might exist was still an open question. Could the proton have its own antiparticle? And if every particle, including the proton, has an antiparticle, then where is all this antimatter?

Today we know that the proton is composed of quarks, making the production of antiprotons a lot more difficult. In the mid-1950s, four physicists—Owen Chamberlain, Emilio Segrè, Clyde Wiegand, and Thomas Ypsilantis—set up an experiment at the Bevatron accelerator in Berkeley, California, to search for the proton's antiparticle. This new experimental search was guided by an extension or broader interpretation of the Dirac theory of relativistic quantum mechanics, which predicted that every charged particle should have an oppositely charged antiparticle. For their experiment, the physicists placed a target in the positive proton beam of the Bevatron accelerator, then collected into a beam line all the negatively charged particles that were produced by the proton interactions at the target. From these particles, they selected only those of a single momentum to pass through their detectors. They were then able to identify the mass of the particles by a combination of two

The Bevatron accelerator at the University of California along with the major components of the experiment that discovered the antiproton. The proton beam in the accelerator hit a copper target; the negative particles produced were collected and transported in a beam line into the time-of-flight and Cherenkov mass identification equipment that distinguished antiprotons from pions and electrons.

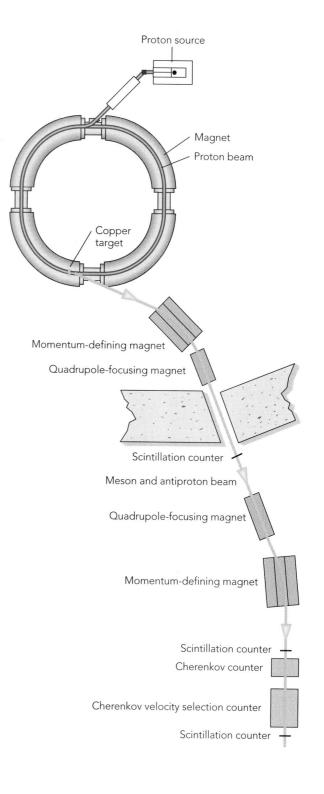

Proton source

Magnet

Proton beam

Copper target

Momentum-defining magnet

Quadrupole-focusing magnet

Scintillation counter

Meson and antiproton beam

Quadrupole-focusing magnet

Momentum-defining magnet

Scintillation counter

Cherenkov counter

Cherenkov velocity selection counter

Scintillation counter

techniques: measuring a particle's time-of-flight in traversing the 30-foot-long beam line, and observing the newly discovered Cherenkov light cones of particles in the beam line.

Particles with different mass but the same momentum take different lengths of time to travel a given distance. Electrons and other low-mass particles would be the fastest; pions, which are heavier, would take longer; and antiprotons, if they existed, would take even longer. Thus by measuring a particle's time-of-flight, the Berkeley physicists could make an estimate of mass. For a second estimate they turned to another method of measuring particle velocity—observing Cherenkov light cones. Cherenkov light, named after the Russian scientist who discovered it, is produced at different angles depending upon the velocity of the particle. Measurement of this angle, combined with the time-of-flight technique, allowed the experimenters to determine a particle's actual mass. This double identification method built more certainty into the experiment. The experiment produced a large sample of data, so that the scientists would not have to rely on just a few events, as in cloud chamber experiments, and as a consequence their results were more reliable.

This new experiment confirmed the existence of antiprotons, for among the negatively charged particles were some with the exact mass of the proton. The notation for an antiproton is the notation for the proton, p, but with the addition of a bar over the top, $\bar{p}$. In general, any antiparticle can be denoted by placing a bar over the notation for the normal particle. However, for historical reasons there are exceptions to this rule. The positron is written as e^+, while the electron is denoted as e^-.

Because the neutron is now known to be composed of charged quarks, it, too, must have an antiparticle. Antimatter therefore is composed of antiprotons and antineutrons, forming atomic nuclei orbited by positrons. The quarks that compose the antiprotons and antineutrons are "anti" as well. For every type of quark there is an antiquark, which behaves in a manner identical to its quark counterpart, but carries opposite quantum numbers such as charge.

The Dirac theory of relativistic quantum mechanics could not distinguish any difference between photons that interact with electrons and photons that interact with positrons. The photon, as the carrier of the electromagnetic force, is only a quantum of energy and has no matter-antimatter distinction. In contrast, the W particle of weak nuclear interactions does come in positive and negative charge states, but both these charge states of the W can interact with either matter or antimatter just like the photon.

Before the Bevatron experiment had identified the antiproton, many cosmic ray experiments had produced pictures of particle interaction events that showed a brilliant star-ray pattern. These events could not

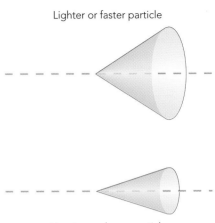

Lighter or faster particle

Heavier or slower particle

Cherenkov light cones produced for particles of differing mass but the same momentum. The opening angle allows physicists to identify what mass particle is going through a detector.

The team of scientists that first observed the antiproton at the Bevatron accelerator of the University of California, Berkeley. On the left is Emilio Sergrè and on the right is Thomas Ypsilantis; behind them stand the racks of electronics that recorded the particles.

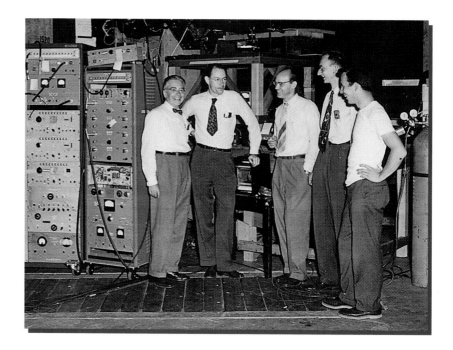

be understood at the time because the older photographic emulsion and cloud chamber experiments provided no precise techniques for particle identification and momentum selection. Today, we know that these star-ray patterns indicate that antimatter particles, such as antiprotons or antineutrons, have collided with their normal matter counterparts and annihilated. When a particle collides with its antiparticle, both particles cease to exist; their annihilation releases pure energy. That ordinary matter is so plentiful in our universe tells us that there cannot be much antimatter, or both forms would still be annihilating each other. The question of why the universe is composed only of matter and not anti-matter will be addressed in Chapter 10.

Many science fiction writers have used matter-antimatter energy re-actors in their novels. These writers know that confining large quanti-ties of antimatter in a magnetic field is probably the most compact means of storing the kind of immense energy needed to send a spaceship from star system to star system in an exploration of space. Our present technical limitations prevent us from building such an energy-storage device, but our ability to produce antimatter is like an infant's first step in this direction. Today, some experiments make intense beams of pure antiprotons and collide them with an accelerated beam of protons. The total annihilation of the colliding particles releases an immense amount of energy, sufficient to produce a large number of new particles to study.

Chapters 7 and 9 will give more details of experiments that use these techniques.

Majorana's Neutrino

Immediately after Enrico Fermi published his theory of beta decay in 1934, a close scientific associate and friend of his, Ettore Majorana, proposed that the neutrino may be its own antiparticle. Since the neutrino and the neutron have neutral charge, they were the only particles that could possibly be their own antiparticles. We now know that the neutron is composed of charged quarks, so this argument no longer holds for that particle. The astute reader might think that because the mesons are made of quark-antiquark pairs, they too should be considered their own antiparticle; however, because the meson is composed of quarks, it is no longer considered elementary. An antiparticle is not just the opposite-charged particle of equal mass. It also has completely opposite spin, and all associated quantum numbers, such as strangeness, are opposite.

Even today, the question of whether the neutrino could be its own antiparticle is still being considered. If the neutrino is its own antiparticle, then it would not always behave as a Dirac-type particle, as prescribed by Dirac's equation. Instead, it would have some additional and unique properties.

After thinking about this possibility, Majorana formulated a theoretical description of the neutrino such that it would be its own antiparticle. According to this special description, an antineutrino produced by the beta decay of antimatter would be indistinguishable from a neutrino produced by the beta decay of ordinary matter. In whatever interaction a neutrino is involved, it would be possible to replace the neutrino by an antineutrino without affecting the process. However, for a neutrino to be its own antiparticle two principles of physics must be violated. First, lepton number conservation would be violated when the neutrino's lepton number changes from +1 to −1 in the process of becoming an antineutrino. Second, the neutrino's spin would not be the same as the antineutrino's spin: the neutrino's universal left-handed spin would change to a universal right-handed spin. However, because lepton number conservation is a principle of purely empirical origin, we may be able to discount it in this case if the spin problem can be worked out.

If it is true that the neutrino has a nonzero mass, there would be a *rest frame* from which it could appear to change helicity and interact as though it had the opposite-handed spin; it would thus act like its opposite particle. This may sound complicated, but with the help of a simple example it should become clearer. Imagine two vehicles traveling in the same direction down a road, one a sports car and the other a truck.

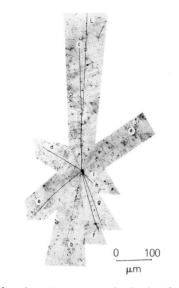

Before the antiproton was clearly identified to exist, many cosmic-ray emulsion experiments saw starburst patterns like the one in this exposure made at the Bevatron. Here an antiproton (track L) produced by the Bevatron has been annihilated in a collision with normal matter, producing pions (tracks a and b) as well as protons and alpha particles (other tracks).

Ettore Majorana, the theoretical physicist who proposed that the neutrino could be its own antiparticle. This possibility continues to haunt us today.

Suppose the sports car going 65 mph passes the truck going 55 mph; how an observer perceives this scenario depends upon his reference frame. From the point of view of an observer in the truck, the sports car passes the truck at 10 mph. But, from the point of view of an observer in the sports car, the truck appears to be moving backward at 10 mph as it is passed. Imagine yourself inside a car: you cannot really tell if it is moving with respect to the ground, but you can see other cars passing. A car doing 10 mph as it passes by you when your car is stopped looks the same to you as a car doing 65 mph passing you when your car is doing 55 mph. Similarly, when you pass a slower car, it appears to be moving backward. You see an *apparent* change in direction.

Suppose, instead of cars, we talk about two spinning particles passing each other. These particles can move in one of either two apparent directions, depending upon the reference frame from which one chooses to observe them. This change in observed direction gives rise to a change of *observed* helicity. As described in Chapter 3, the direction of spin is always relative to the direction of motion; change the direction of motion and you've changed the helicity, or spin handedness, about the direction of motion. But for particle interactions, the only reference frame of any importance is the one in which particles interact, not the reference frame of a noninteracting observer. In the theory of relativity, a particle traveling at the speed of light will appear to be going at the speed of light no matter what the reference frame, but remember that only a massless particle can do that. A particle that has nonzero mass can never travel at the speed of light, however small that mass is; it is always possible to find a slightly faster rest frame. Therefore, we can safely conclude that it is in the realm of possibility for a neutrino of nonzero mass to change to opposite-handed spin, although it is not possible for a neutrino of zero mass. This brings us back to the question of whether the neutrino has mass, first discussed in Chapter 2. If it does, then there is the possibility that the Majorana theory is correct and that the neutrino is its own antiparticle. And if that is the case, then some nuclei could undergo a special form of double beta decay.

Double beta decay occurs when a nucleus undergoes two beta decays simultaneously, or at least as simultaneously as the Heisenberg uncertainty principle requires. In ordinary double beta decay, as might be expected, a nucleus emits two electrons and two neutrinos. But in another type of hypothesized double beta decay, the neutrino emitted in the first decay is reabsorbed in the second decay as the antiparticle. This type of decay is only possible if the neutrino is of the Majorana type. The neutrino would have to have changed spin states, and such a change is only possible if the neutrino has a nonzero mass. No neutrinos are emitted from the process for observation, but two electrons are, and there is a resulting nucleus that has been transformed to an element with two additional protons. If scientists were to observe such a decay, it would tell us that the neutrino was a very special particle indeed, one that can change lepton number from −1 to +1, and one that can also be its own antiparticle, something no other elementary particle can be. It would also definitely tell us that the neutrino's mass is nonzero.

The nucleus of an atom of a particular element may decay into the nucleus of another element through the radioactive processes discussed in Chapter 1. However, energy conservation forbids the resulting nucleus from being heavier than the initial nucleus. Yet elements exist that can have two simultaneous beta decays even though a single beta decay might be forbidden. Such elements would be perfectly suited for use in a search for double beta decay. Some nuclei that are stable except through the process of double beta decay are:

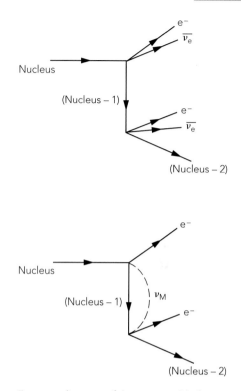

$$^{76}\text{Ge} \rightarrow {}^{76}\text{Se} + e^- + e^- \qquad {}^{136}\text{Xe} \rightarrow {}^{136}\text{Ba} + e^- + e^-$$

$$^{82}\text{Se} \rightarrow {}^{82}\text{Kr} + e^- + e^- \qquad {}^{238}\text{U} \rightarrow {}^{238}\text{Pu} + e^- + e^-$$

This list shows us only a few of the many possibilities.

Scientists have devised two methods for observing double beta decay: direct observation of an event or the measurement of a physical parameter such as electron-energy distribution or geophysical isotope ratio, both indirect techniques.

The search for direct evidence of double beta decays began during the latter half of the 1930s and continues to the present day. Such a search may proceed by one of two means. First, physicists can observe directly the two electrons produced by double beta decay events. The problem is how to tell whether the electrons come from the same nucleus at the same time instead of from two separate but nearby nuclei undergoing single beta decays. It would be most advantageous if the two electrons from the double beta decay could be observed emanating from the same place simultaneously. Alternatively, physicists can measure the electron-energy spectrum from double beta decay. The electron-energy distribution of regular double beta decays would be distinctly different from that of single beta decays, and the electron-energy distribution of neutrinoless double beta decays would in turn be distinctly different from that of regular double beta decays. The reason is that in neutrinoless double beta decay, unlike in regular double beta decay, the energy of the electrons would equal the mass difference between the initial nuclei and the daughter nuclei. This second technique allows physicists to use a higher-density detector, able to detect a greater number of double beta decays, while the first is more appealing to the perfectionist who might wish to see a handful of decays as they take place. Both techniques face the problem of distinguishing true double beta decays

Feynman diagrams of the two possible forms of double beta decay. In the lower form the neutrino of the first decay is reabsorbed as a neutrino of opposite spin in the second decay. Hence the second decay emits no neutrinos. This form of beta decay can occur only if the neutrino can be its own antiparticle.

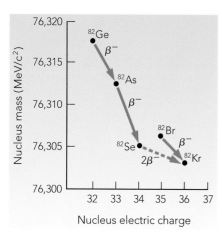

The elements plotted on this graph all have 82 protons and neutrons in their nuclei but differ in the number of protons and thus in mass and nucleus electric charge. The nuclei of selenium-82 cannot undergo beta decay (symbolized β^-) to form bromine-82 without violating the energy conservation law, but it can undergo a double beta decay to krypton-82 by passing through a virtual bromine state within the limits of the Heisenberg uncertainty principle.

from a large background "noise" consisting of random electrons emitted by other natural processes occurring simultaneously in the detector.

The first observation of regular double beta decay, the type emitting two neutrinos, resulted from a chain of experiments conducted over more than a decade by M. K. Moe and colleagues at the University of California at Irvine. In 1987 they reported a direct observation of double beta decay in selenium-82 (^{82}Se), much to the excitement of the physics community. The race began to be the first to observe neutrinoless double beta decay.

A successful search for neutrinoless double beta decay would prove that the neutrino could be its own antiparticle, as Majorana suggested, and that the neutrino was not massless. A flurry of experiments since 1987 have confirmed that several elements undergo double beta decay in the two-neutrino emission mode, but an observation of the neutrinoless double beta decay remains elusive.

The second technique used to search for double beta decay comes from the field of geology, and in fact the results provided by geologists are responsible to a large degree for our understanding of the physics up to this point in the story. The technique is similar to carbon dating, the means by which archeologists determine the age of animal, human, or plant remains. For example, when an animal dies, its carbon intake stops and its carbon-14, which is naturally radioactive, continues to decay into carbon-12. An archeologist can calculate the approximate age of an animal's remains by determining the ratio of carbon-14 to carbon-12 in those remains and figuring the known half-life of carbon-14 into an equation.

Geologists are able to use a similar technique to help estimate the half-lives of isotopes that undergo double beta decay. They find an old

The electron-energy spectrum observed from normal beta decay, and the expected distribution of energy carried by the electrons of double beta decay, for the two types of such decay. Neutrinoless double beta decay is clearly distinguished from the other beta decays by the fact that the electrons it releases always carry more or less the same energy, the result of not having to share energy in differing amounts with another particle.

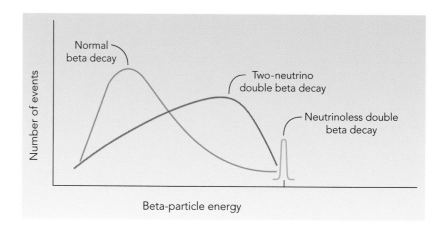

rock deposit rich in elements that undergo that type of beta decay: this deposit should be stable and undisturbed by nature for a known amount of time. The geologists then analyze the deposit's isotope contents. Geological dating techniques provide an estimate of the age of the trapped rock formation. A comparison of the rock's age with the measured ratio of daughter to parent isotopes gives a measure of the rate of double beta decay. Unfortunately, this technique does not distinguish between the two types of double beta decay. However, as long as scientists have a direct measure of one type of double beta decay, they can deduce limits on the other type. If the rate of regular double beta decay is not sufficient to account for all the transformed elements present in the sample, then some other form of double beta decay could still be present.

Historically, geophysical limits on double beta decay have played a crucial role in keeping interest in this field alive. It was the geologists who, by providing ample evidence that double beta decay should exist, encouraged physicists to keep looking. After Moe's experiment directly observed double beta decay in 1987, the geologists continued—and continue now—to help guide the field. They have pointed out a discrepancy between their half-life measurements of double beta decay and the direct observation value that tells us that there still may be another double beta decay mode. This discrepancy is just the encouragement that physicists need to keep looking for neutrinoless double beta decay.

A new technique involving radioactive-isotope counting and precise chemistry, called radiochemistry, is providing additional hints that the search may one day prove fruitful. In 1991, Tony Turkevich at the University of Chicago and collaborators found a signal in their analysis of thirty-year-old uranium-238. This isotope of uranium cannot undergo single beta decay, but it can undergo double beta decay into plutonium. The plutonium isotope formed is radioactive and can be precisely measured, even in small concentrations, by radioactive counting. Although this technique does not provide direct evidence for neutrinoless double beta decay, an interpretation of the results does suggest that this form of double beta decay may be occurring. A possible flaw in the experiment, which makes the results controversial, is that the uranium may have been contaminated. The group had used old samples of uranium left over as a weapons by-product, and these samples may not have been properly isolated from naturally occurring radioactive processes. Nevertheless, the results are stimulating and provide further encouragement for continuing the search.

Experiments to detect double beta decay directly have been improving in the 1990s, giving hope that the detection of a Majorana-type neutrino and neutrinoless double beta decay may be within reach. The most sensitive experiment to date is the Heidelberg-Moscow collaboration, which is located deep beneath the Apennine Mountains in the Grand

The setup of the Heidelberg-Moscow double beta-decay experiment. The innermost cylinder is a sample of germanium rich in ^{76}Ge, one of the nuclei that can undergo double beta decay. In this experiment, the germanium is both the element being studied for double beta decay and the scintillator that reacts to the electrons being released, emitting detectable flashes of light. The high-purity copper surrounding the experiment shields the sample from naturally occurring background radiation and cosmic rays.

Sasso laboratory. This experiment is searching for neutrinoless double beta decay in the element germanium-76 (^{76}Ge). The scientists involved are measuring the total energy of the electrons emitted in regular double beta decay in the hope that they will find evidence of events in which the emitted electrons get all the energy. If they are successful, then they can see whether there is a way to enhance these events so as to obtain evidence for neutrinoless double beta decays. This laboratory has the best half-life mass limits on neutrinoless double beta decay. When interpreted, the results give a mass limit of less than 1.1 eV/c^2 for the Majorana neutrino. This limit on the neutrino's mass would apply only if the particle were shown to be its own antiparticle.

Neutrino Oscillations

Chapter 5 briefly discussed the discovery of the quark composition of matter. This theory, which explains nuclear matter as protons, pions, and other particles composed of quarks that interact with the strong nuclear force, has led to a new organization of matter. However, I certainly did not explain everything we currently know about this organization back in Chapter 5. One thing I left out is that the neutrino may possess a special property of matter that permits it to transform itself from one type of neutrino into another.

This possibility is perhaps even more bizarre than the possibility that the neutrino may be its own antiparticle. Both possibilities require that the neutrino have a nonzero mass. In addition, the ability of one type of neutrino to act like other types of neutrinos and the neutrino's ability to be its own antiparticle would both be manifestations of the same basic property of the neutrino, an ability to transform itself. A neutrino with this property would fit into this world or the antimatter world just as comfortably.

The many particles discovered in the golden age of elementary particle physics of the late 1940s exhibit two distinctly different lifetimes. Some particles, such as the delta particle (Δ), are unstable with respect to the strong nuclear force. These tend to decay very quickly, and have a mean lifetime of only 10^{-23} second or even less. Other particles live longer, with a typical mean lifetime of 10^{-8} second. These longer-lived particles have an unusual property: they violate conservation of strangeness, one of the new quark quantum numbers, but only through the weak nuclear force decay. Two examples are the decays of the lambda and negative kaon particles:

$$\Lambda^0 \rightarrow p + e^- + \overline{\nu}_e$$

$$K^- \rightarrow \pi^0 + e^- + \overline{\nu}_e$$

These weak decays appear almost like normal beta decays, except that, instead of a down quark coupling to a W and producing an up quark in the final state, a strange quark couples to a W and produces an up quark. Everything else in the decay process functions exactly the same. It is as if the strange quark really thinks itself to be a down quark and acts accordingly. In the process it violates strangeness conservation, but it is only through the weak-force interaction that strangeness, or any other quark quantum number, is violated.

In 1963 the theorist N. Cabibbo put forth an explanation for this phenomenon. His theory is essentially a selection rule, which states that the change in the strange quantum number must be equal to the change in electric charge of the primary particle when strangeness is not conserved. This simple rule summed up all the known strangeness-violating decays. Cabibbo's theory also went a step further and provided an explanation of this behavior. The down and strange quarks are not pure states, the theory explains, but are instead a mixture of states. This property of down and strange quarks is indicated by the prime symbol,

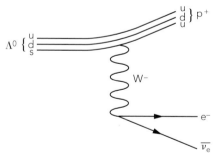

A Feynman diagram of lambda beta decay at the quark level.

d′ and s′, in the table of quarks and leptons found in Chapter 5.* The mathematical formulation of the theory tells us that d′ is mostly just a down quark that has a small part that behaves like a strange quark. Similarly, most of the time s′ is a strange quark, but there is a small part that will act as a down quark instead.

If it were not for the mixing property of the down and strange quarks, many of the new particles would actually be stable. We could then have had exotic stable matter much heavier than is normally seen. The mixing property also explains the somewhat longer lifetimes of the particles that violate strangeness conservation. These particles are stable with respect to the strong force, and would not decay at all if mixing did not eventually cause their decay through the weak force. Their lifetimes are longer than those of particles that are unstable with respect to the strong force because they must wait longer before the weak force will permit them to decay.

Since only one set of quarks mix in weak interactions, while the up and charm quarks do not exhibit any mixing of this nature, a logical question to ask is whether the leptons also have a set of particles that mix. The charged leptons—electrons and muons—seem not to mix. This leaves only the neutral leptons—the neutrinos—to be examined for possible mixing. The first experiment to search for neutrino mixing, now referred to as neutrino oscillations, was carried out at CERN in Geneva, Switzerland, in the early 1980s using the CDHS detector.

In the experiment, a low-energy neutrino beam from the 26 GeV/c² Proton Synchrotron (PS) accelerator passed through two detectors. One detector was located close to the origin of the neutrino beam at 130 meters (440 feet) and the other farther from the origin at 885 meters (half a mile). Physicists compared the neutrino beam at the two locations, looking for any changes that neutrino oscillations could have caused while the beam traveled from the near site to the far site. They were able to search for many different types of neutrino oscillations, including flavor changes such as electron-type neutrinos changing to muon-type neutrinos and vice versa. Neutrino oscillations are similar to the Cabibbo mixing already seen in the down and strange quarks, yet they are also different because the neutrinos can change back and forth with

---

*In more detail, Cabibbo's theory says that the observed down quark labeled d′ in the (u, d′) doublet that couples to the weak-force propagator particle is actually

$$d' = (d \cos \vartheta_c + s \sin \vartheta_c)$$

and that the observed strange quark labeled s′ is also a mixture:

$$s' = (s \cos \vartheta_c + d \sin (\pi + \vartheta_c)),$$

where ϑ_c is a small mixing angle so that $\cos \vartheta_c = 0.9$ is almost 1, but not quite. Similarly, $\sin \vartheta_c$ is closer to zero but not exactly equal to zero.

The CDHS experiment at CERN, which made one of the first searches for neutrino oscillations. The many layers of detectors are separated by large slabs of steel; these slabs provide a high density of matter to help stop some of the neutrinos.

time. Moreover, neutrino oscillations are affected by the matter through which the neutrino propagates; the more matter, the higher the likelihood of oscillation. Physicists can even search for a neutrino oscillation that would convert a neutrino into its antiparticle, a transformation that would change the lepton number from -1 to $+1$ as predicted by Majorana's neutrino theory.

This first experiment actually found evidence supporting neutrino oscillation. It captured the attention of the physics and even astronomy communities, since evidence that neutrinos can change from one flavor to another as they pass through matter in space could help explain the missing solar neutrino problem. Because the solar neutrinos are produced in the core of the Sun, they must traverse all the matter in that star before traveling through space to reach the Earth. A small effect over such a large distance could result in a large neutrino oscillation from one flavor to another—in this case, an oscillation from electron-type to muon-type neutrinos. If large enough, such an oscillation would account for the missing solar neutrinos. Since the detectors are insensitive to muon-type neutrinos, the transformation of some of the electron-type neutrinos into muon-type neutrinos would make there appear to be fewer neutrinos from solar nuclear reactions.

This flurry of activity in neutrino-oscillation physics was not to last; no other experiment was able to confirm the first experiment's report, even though the many experiments that followed were more sensitive and controlled background noise better. If the reported effect had been

The Los Alamos experiment of 1995, which may actually be seeing the first signs of neutrino oscillations, uses this large tank to detect the neutrinos in a particle beam. Although shown here empty, the tank is usually filled with a special mineral oil laced with scintillating chemicals. Its walls are lined with photosensitive detectors to record the faint track of light created when a neutrino interacts with an atom in the liquid.

real in the first experiment, then these subsequent experiments should have been able to see it all the more clearly. The fact that they did not means that neutrino oscillation is still only a theoretical speculation.

The failure to find neutrino oscillations in subsequent experiments provides us with another good example of how science works. We have already seen similar cases illustrating that experimental results must be confirmed in order to be credible. When making measurements to search for a small effect, scientists must be aware of many sources of error, and each source must be fully eliminated, controlled, or at least estimated before an experiment can be performed at the necessary level of sensitivity. At present, most physicists would agree that CERN makes the most accurate measurements in the field of particle physics, but even with the laboratory's good reputation for attention to detail and care in performing experiments, its results must be scrutinized.

Other physicists have gone as far as writing books and lectures about the many blunders in physics. Although these stories are somewhat amusing, especially to physicists, I would not suggest contemplating them too deeply. On publishing their first result the scientists who discovered the antiproton accidentally put the wrong time axis on their time-of-flight data; as a result the antiproton appeared to be going faster than the speed of light. According to Tom Ypsilantis, this small mistake occasionally brought letters to Emilio Segrè from science fiction fans who wanted to know why scientists were suppressing the discovery

A proposed long-baseline neutrino-oscillation experiment will send a beam of neutrinos from Fermilab, near Chicago, to the Soudan mine in northern Minnesota.

of a faster-than-light particle. One should never be ashamed of a mistake, unless it is committed intentionally. It is through mistakes, or by searching for what theory tells us cannot be, that our understanding of any field of science grows.

Today, the search for neutrino oscillations continues, but no experiment at present claims to have definitive proof that the phenomenon occurs. Possibly, the experiment that has come closest to seeing neutrino oscillations is being performed at Los Alamos National Laboratory; it may or may not have found a hint of a signal. The data that the Los Alamos experimenters have taken shows a much smaller level of neutrino oscillation than the data from the initial experiment at CERN, and they are not ready to claim a certain discovery. If their observations of the level of neutrino oscillation are correct, then the mass of the neutrino is further constrained to be less than 0.2 eV/c^2, an extremely small value. The Los Alamos laboratory has the best accelerator for this type of precision experiment because of its extremely intense and clean primary particle beam. Perhaps in the next few years the experimenters at Los Alamos will be able to take sufficient data to clearly identify neutrino oscillations.

The next round of experiments being proposed, which would have an even better sensitivity to neutrino oscillations, could make use of the accelerator at the Fermi National Accelerator Laboratory, or Fermilab for short, in Batavia, Illinois, near Chicago. The accelerator would aim a beam of neutrinos deep underground to a mine in Soudan, Minnesota, to create the longest distance man-made neutrino-oscillation experiment ever performed, covering 450 miles and reaching a depth of six miles. The hope is that increasing the distance between the two points being compared will improve the accuracy of the measurement.

Two other experiments searching for short-length neutrino oscillations are at present underway at CERN. None of these experiments are certain to achieve positive results, but even if no experiment ever sees the theoretically proposed neutrino-oscillation effect, the limits they set are an important guide for scientists who study neutrinos that come directly from the core of the Sun. Eventually, accelerator experiments may rule out neutrino oscillations as a possible explanation for the missing solar neutrinos, forcing scientists to search by other means for an answer to this mystery.

· · ·

A peculiar mixing of the down and strange quarks has inspired physicists to speculate that neutrinos may undergo a similar mixing. We are still uncertain whether the effect is real, and are likely to continue so until additional experiments are carried out to look not only for mixing between the different flavors of neutrinos, but also for a type of neutrino that is its own antiparticle. Amazingly, a successful observation of neutrino oscillations would also confirm the existence of the Majorana neutrino.

After Fermi's theory of 1934 resolved the question of the neutrino's actual existence, Majorana asked the question that is still haunting particle physics today: Can the neutrino be its own antiparticle? The answer to this simple but unresolved question could also tell us whether the neutrino has mass. However, the converse is not true; if Majorana-type behavior is never found, the neutrino may have mass anyway. If the neutrino is ever found to be of the Majorana type, the discovery would add to its already perplexing character and make it truly one of the most mysterious particles of nature.

A proton-antiproton collision at CERN's Super Proton Synchrotron produces a spray of particles and antiparticles. Among the particles, this image catches a highly energetic electron, highlighted by a pink arrow. This electron was the signature of the first observed production of a W particle.

CHAPTER 7

Toward a Unified Theory of Electro-Weak Interactions

Scientists had known about weak interactions and the neutrino for about forty years when, in 1973, a novel experiment surprised the physics community by revealing a new type of neutrino interaction. This discovery, so long after the neutrino had first been conceived, became a pivotal turning point in the story of weak interactions. It would trigger a flurry of new theoretical formulations that finally provided a framework for understanding weak interactions. Even more important, these new theoretical formulations showed that weak interactions are actually part of a grander theory, one that also encompasses the interactions of photons with matter.

Neutral-Current Neutrino Interactions

Although everything about Yukawa's theory of weak interactions seems just fine on the surface, I was not completely honest back in Chapter 3. His theory was undeniably elegant, and it was able to explain the many newly discovered decays involving particles such as the pion and the muon, yet in reality this new theory was more incorrect than Fermi's four-point beta decay. Fermi's simpler four-point theory failed completely once the zoo of new particles had been discovered, but it matched experimental results for simple radioactive beta decay much more accurately. During his lifetime, Fermi had acquired a reputation for always being right. Since his death, during this struggle to understand why the new formulation of weak interactions did not predict measured parameters accurately, Fermi acquired a new reputation: even when he was wrong he still managed to be right!

Something was certainly wrong with Yukawa's theory, and it could not be fixed by using the new W weak-force propagator particle. The fact that theorists were having trouble with Yukawa's theory hinted that something critical to a full understanding of weak-interaction physics had yet to be understood.

Two technicians working inside the gold-plated interior of Gargamelle, the heavy-liquid bubble chamber that first found neutral-current neutrino interactions.

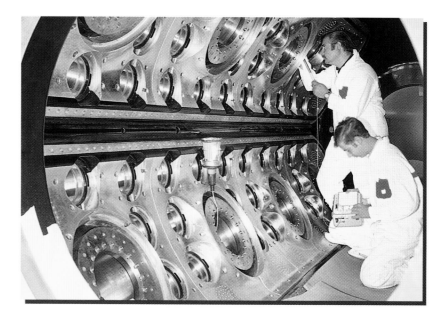

In the early 1970s, the European Center for Nuclear Research (CERN) began a new experiment using the Proton Synchrotron accelerator, which was of a somewhat lower energy than the Brookhaven AGS machine, and a heavy-liquid bubble chamber called Gargamelle. The experiment—an international collaboration between France, England, Italy, Belgium, and Germany—was led by André Lagarrigue of Orsay, France. Other famous physicists—Paul Musset and A. Rousset of CERN and Don Perkins of Oxford, England—led various groups in the taking and analysis of data from Gargamelle. This unique and challenging experiment would eventually provide the crucial piece of the weak interaction puzzle, a piece that would guide theoretical physicists to the first unified theory of weak interactions and electromagnetism.

The new experiment differed from others that had come before it in that the bubble chamber was filled with a very heavy liquid, Freon-13, instead of the lighter and more typical liquid hydrogen. In this higher-density liquid, more neutrino interactions would take place per burst from the neutrino beam, and pions were more easily distinguished from muons. In addition, unlike conventional bubble chambers, which are cylinders that typically stand upright, Gargamelle was a round cylinder that lay on its side. This arrangement provided more length along the beam direction, which in turn provided an opportunity for many more neutrino interactions to occur. Moreover, the large bubble chamber could fully contain the particles produced from a single interaction. The neutrinos entered the chamber in a nearly pure beam of muon-type antineutrinos, although the possibility existed that the beam was slightly contaminated with electron-type antineutrinos. During its lifetime, the Gargamelle experiment took almost a million pictures of neutrino interactions, capturing multiple interactions in each individual picture frame. Each picture was meticulously scanned onto a projection table and carefully searched for interesting events.

A typical neutrino interaction observed in previous experiments began with a neutrino entering the chamber and interacting with a proton or neutron through the exchange of a heavy virtual W particle. The interaction produced a charged lepton, such as an electron or muon, along with the baryon that the neutrino had interacted with. The Gargamelle experiment was actually started as an effort to physically produce real W particles and measure their mass. Measuring the W's mass would help theorists find the correct interaction strength of the weak force, not to mention the fact that it would prove that the W really existed. However, the Gargamelle experiment was soon to discover something even more incredible than a real W particle. The CERN physicist Paul Musset led an analysis showing that many events produced no charged leptons at all. The scientists were baffled, since it should be impossible for the charged W particle to mediate an interac-

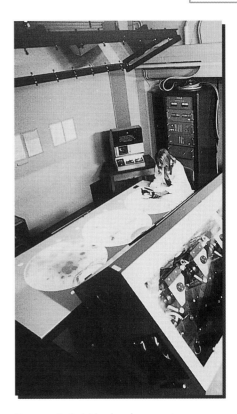

Gargamelle bubble chamber events recorded on film were projected on this scanning table. Using rulers and compasses, scientists then took measurements from the projected photos.

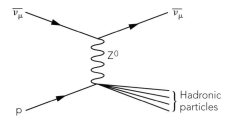

A Feynman diagram of the neutral-current weak interaction of the hadronic type; no charged leptons are emitted from the interaction. These events proved that a neutral propagator of the weak nuclear force must exist.

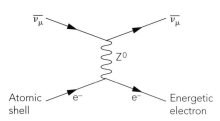

A Feynman diagram of the neutral-current weak interaction with an atomic-shell electron. Evidence for this interaction helped convince physicists of the existence of the Z^0 particle.

tion that did not produce a charged lepton. Was it possible that another weak-force propagator particle existed, this one of neutral charge?

It was postulated that these events were what are called neutral-current neutrino interactions. In this type of interaction a neutrino enters the detector and interacts with a neutron or proton by the exchange of a chargeless weak-force propagator particle, the Z^0. Hence the name "neutral current"; "charged-current" interactions, in contrast, are those mediated by the positive and negative W particles. A neutrino also comes out of the interaction, conserving lepton number, while additional energy is imparted to the neutron or proton that was involved, causing it to interact with other hadrons that happen to be along its flight path.

The Gargamelle experiment recorded many such events distinguished by an absence of leptons. If these events were indeed the work of a neutral weak-force propagator particle, then physicists expected to observe another type of neutral-current neutrino interaction in which a neutrino would interact with an atomic-shell electron. An analysis overseen by Don Perkins provided just such an event. Part of his group, working in the West German town of Aachen, found the first event wherein a muon-type antineutrino scattered off an atomic-shell electron. The antineutrino imparted some of its energy to that electron, which, liberated from the atom, followed a path that the scientists could trace in a bubble chamber.

Perkins tries to convince the English-speaking world through his textbook *Introduction to High Energy Physics* that, because these events are so clean, the observation of neutral-current neutrino interactions with atomic-shell electrons was the turning point toward proving neutral currents. Charles Peyrou, a CERN division leader at the time of the Gargamelle experiment, has correctly pointed out the flaw in this argument. Although the observation of an atomic-shell electron made energetic through a neutral-current interaction was a good way to double-check the original result, the observation would have been useless without the more copious events observed in Musset's analysis, since the muon-type antineutrino beam was known to be contaminated with a small number of electron-type antineutrinos. An electron-type antineutrino could have produced electrons by interacting through the charged-current exchange of a W particle. Thus a single or a few events producing only an electron could have easily been attributed to some contamination in the muon-type neutrino beam line.

In July of 1973, the CERN Gargamelle experimenters officially announced the discovery of neutral-current neutrino interactions. This discovery was a significant turning point in the development of theoretical particle physics; in fact, it is considered to be the dividing line between modern theories of particle interactions and the older, classical theories developed when particle physics was in its infancy. The results were pre-

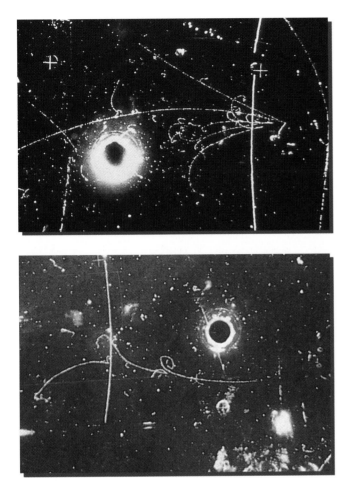

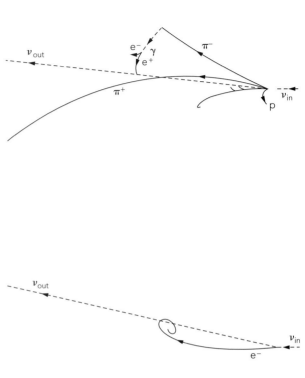

sented at a conference attended by the University of Chicago physicist Val Telegdi, mentioned in Chapter 3, and his student Roland Winston. When asked by Winston whether the results were important, Telegdi replied that this was the most important discovery of the decade.

Many famous theorists had predicted the existence of a weak interaction involving a neutral, force-propagator particle. Theoretical physicists tend to be better remembered than experimental physicists because a theory bears the name of the theorist and these theories live on in textbooks long after the experiments that confirmed the result are forgotten. In this case, however, although these theorists had made a prediction that was very much correct, many physicists had never heard of it, probably because theorists are so prolific in producing theories that it is difficult to keep up with them all. For almost any experimental result, a theorist has probably already speculated on the possibility of its exis-

Two images of neutrino interactions from the Gargamelle bubble chamber, accompanied by sketches identifying the particles. In both images, the neutrino enters from the right. The top picture with its many curling particles shows a neutral-current hadronic interaction; no charged lepton is emitted from the interaction. The bottom picture, having only one small curling particle, shows a rare neutral-current interaction with an atomic-shell electron.

Experimentalist Paul Musset (right) and theorist Abdus Salam toasting the discovery of neutral-current neutrino interactions. Both men were crucial participants in the work toward an understanding of neutrino interactions. Their toast is symbolic of how experiments must agree with theory. (Abdus Salam, as a Muslim, has a glass of orange juice.)

tence. In this case, I consider the triumph an experimental one because the first observation of a neutral weak-force process helped guide the way for theory. If the Gargamelle experiment had not seen neutral-current interactions, these papers would most likely still be unknown.

Other Opportunities to Observe Neutral-Current Interactions

Historically, the first experiment to search explicitly for neutral-current neutrino interactions had been led by Pontecorvo at the Dubna accelerator, 100 miles north of Moscow, in 1961. This experiment, along with similar experiments at CERN, Brookhaven, and the Argonne National Laboratory, was begun long before the CERN researchers started looking for such interactions using Gargamelle. These experiments all had the potential to discover the neutral-current neutrino interactions. Why they did not provides another insight into how science is conducted.

One of the leaders of the Argonne experiment, Malcolm Derrick, has explained how his group missed out on being the first to observe neutral-current neutrino interactions. In 1963, Derrick and his group conducted an experiment similar to the Gargamelle experiment but using the 12 GeV Zero-Gradient Synchrotron (ZGS) accelerator and a 12-foot bubble chamber. Although the ZGS machine had a lower energy than CERN's accelerator, the neutrino beam was copious enough that the physicists should have been able to see neutral-current events. But Derrick had taken a class on weak interactions in particle physics at Carnegie Tech, now Carnegie Mellon University of Pittsburgh, where he was taught that neutral-current weak interactions did not exist. The lecturer was Lincoln Wolfenstein, who had been a prominent theorist of particle interactions since the 1950s. During his lectures, Wolfenstein wrote out all the possible weak interactions on the blackboard, then proceeded to cross off all the neutral-current interactions, explaining that such interactions cannot exist! Because of this skepticism on the part of Wolfenstein and other theorists at the time, including Feynman, Derrick's group did not make a thorough search for such neutrino interactions when analyzing the data taken by the ZGS bubble chamber.

From this story it would be fair to conclude that sometimes an experimental scientist must search for phenomena that a theorist says cannot exist, just to confirm that it really doesn't. This is the converse of the lesson illustrated at the beginning of Chapter 6, when Dirac's theory proposing the existence of antimatter helped experimenters find the antiproton and correctly understand the positron observation—allowing physicists to make a clean connection between this actual antimatter and the antimatter of theory. In the case of neutral-current neutrino interactions, the experimenters had to help theoreticians determine which of their models was actually correct. Today, many experimental scientists remember such lessons; searches for exotic processes that are only speculative continue.

Despite its failure to notice neutral-current interactions, the Argonne ZGS neutrino program did manage to do much good. Theirs was the first large bubble chamber experiment to use a neutrino beam to precisely measure the likelihood that neutrinos would interact with other particles at these energies. The Argonne physicists showed that the proton had the same physical dimensions for weak-force interactions as for electromagnetic interactions. Their most celebrated neutrino interaction event came from a detailed study of their data, published in 1974; that event proved that associate production of strange matter in pairs of particles with opposite strangeness is possible, even by neutral-current neutrino interactions. Although these events are rare, a single event of this type from the ZGS bubble chamber was able to confirm that the neutral-current interaction process was real. However, this single event alone could not have constituted the discovery of neutral-current interactions, even if it had been observed before the Gargamelle results were announced.

Shortly afterward physicists at Fermilab conducted an experiment, led by Dave Cline and Carlo Rubbia, that also searched for the reported neutral-current process seen in Gargamelle. This experiment was much smaller, and the data produced by its detector not as well understood. At first, Cline and Rubbia confirmed the neutral-current process,

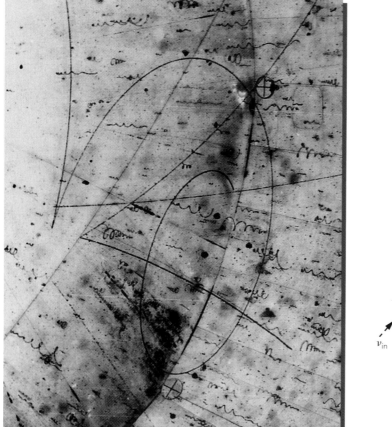

This image recorded in the Argonne bubble chamber shows the production of two strange particles by a neutral-current neutrino interaction. Recognizable only by their decay products, these are the particles labeled Λ^0 and K^0 in the accompanying sketch.

although their events seemed to be occurring at a much lower frequency than reported by Gargamelle. In less than a year, however, they had to announce that the neutral-current signal had gone away and could therefore not be confirmed. The famous CERN physicist B. Gregory, when asked if he believed their announcement that neutral-current interactions no longer existed, replied, "I never believed their original confirmation experiment was correct!"

Naturally, the announcement by the Fermilab experimenters worried the Gargamelle scientists. At that time, the CERN experiment had seen only one superbly clean event produced by an atomic-shell electron interaction, but their sample of neutrino neutral-current interactions that produced no charged leptons was considered large enough to leave little room for doubt. The best data, of course, were from the large, statistically significant sample; the one superbly clean event could not be counted on since it was unique. Adding to the concerns of the CERN group was a discrepancy in measurements of the ratio of neutral-current to charged-current neutrino interactions; their ratio was much higher than the 5% ratio that Fermilab had found in their first result.

Eventually the Fermilab experimenters managed to better understand their detector. They were then able to confirm Gargamelle's result at almost exactly the same level as CERN had originally published. As C. Peyrou said during a discussion, "Fermilab initially saw the effect even though maybe not at the same strength as Gargamelle." As for the Gargamelle experiment, it was ultimately able to produce two more neutral-current neutrino interactions with atomic-shell electrons at the expected rate. By the time the experiment was shut off, it had produced a total of three neutral-current neutrino interactions with an atomic-shell electron, and an even larger sample of neutrino interactions that produced no charged leptons. Although the Gargamelle experimenters, like all good scientists, were justly worried when their results could not be confirmed, in the end their original announcement was proven correct, and neutral-current neutrino interactions were soundly established.

After the existence of neutral-current neutrino interactions had been announced at CERN, physicists went on to compare the number of neutral-current events lacking a charged lepton to the number of charged-current events producing a charged lepton. They found that the muon-type neutrinos interacted a quarter of the time through the neutral-current mode, and the muon-type antineutrinos interacted almost half of the time through the neutral-current mode. Clearly, the neutral-current interactions of neutrinos were not a negligible secondary effect, but a substantial component of weak interactions. It is surprising that such a large effect was not found sooner, but as we've seen, it is sometimes difficult to recognize something experimentally without first having a reason to see it. When CERN physicists finally did notice the effect, they doggedly pursued it until they had an experimental success; according to Peyrou, "It was Musset who took the whole search for neutral currents in his hands." This experimental evidence was a turning point in the field of particle physics, for it would lead the way to a unified theory of electro-weak particle interactions.

In the last chapter, we saw that the strange quark can sometimes act like a down quark for a small fraction of the time. Although this type of behavior is observed in charged-current interactions involving W particle exchanges, no such behavior has ever been observed in neutral-current interactions.

A Unified Theory of Electro-Weak Interactions

The nineteenth-century physicist James Clark Maxwell wrote his doctoral thesis on the unification of the electric force with the magnetic force. In a brilliant stroke of insight, he was able to show that magnetism is only a manifestation of electric charge in motion. His work eventually lead to the four so-called Maxwell equations that describe all the laws of electromagnetism. This is a perfect example of what is meant by unifying two forces into one grand theory. In this case not only were the established laws describing the electric and magnetic forces represented in the new theory, but additional terms in the formula were present that gave more information about this force of nature.

As we saw in the previous section, in the years from 1961 to 1968, several theorists had published papers containing the first detailed theoretical work proposing a neutral particle that could propagate the weak nuclear force. The main theorists who had published such papers were Sheldon Glashow, Abdus Salam, and Steven Weinberg. Each of these men had developed a similar theory independently of one another, and each published his theory without having knowledge of the others. All three theorists had noticed that the then-speculative neutral-current neutrino interaction looks very similar to photon exchange in electromagnetic interactions. There were only two differences: there was an additional charged-current exchange of a W particle in weak interactions, one that was first seen in beta decay, and the mandated left-handed spin interactions of the weak-force W particles had no equivalent in electromagnetic interactions. By formulating the theory of weak interactions in the same format as QED and merging the two together, the theorists formulated the first unified theory that combines both weak nuclear interactions and electromagnetic interactions into the same set of equations. This combined theory is called the theory of electro-weak interactions. Because each physicist's theory is not as fully complete as when all three are combined, Salam, Glashow, and Weinberg are usually given joint credit for the theory of electro-weak interactions.

In electromagnetic interactions, the photon interacts with a particle of positive or negative charge and can produce an electron-positron pair at low energy or a quark-antiquark pair at high energy. Similarly, the neutral Z^0 particle interacts with a particle of left- or right-handed spin and also produces identical leptons of opposite charge, in this case a neutrino-antineutrino pair, or a quark-antiquark pair.

Theoretical physicists have been guided by symmetries of nature in their formulation of new theories. An example of such a symmetry is the behavior of charged particles when their polarity is reversed; a positive particle passing through a magnetic field deflects along a similarly curved path as a negative particle, but in the opposite direction. Another symmetry is observed in neutral-current interactions: both particles with left-handed spin and right-handed spin interact with the neutral weak-force propagator Z^0. In contrast, only particles with left-handed spin interact with the charged weak-force propagator W. To describe this difference between the spin of particles interacting with the

W and Z^0 particles, we say that symmetry has been broken, something that does not appear in the electromagnetic-force interaction of photons. For a common example of symmetry breaking, we can turn to the human body. At first glance, we assume that the right and left halves are symmetric. When we read a newspaper or book, for example, we use both right and left eyes; however, when we write we use only one hand, either right or left. In early childhood children are equally clumsy at handling objects with either hand, but slowly, by their own choice or their parents' insistence, they become either right-handed or left-handed writers and the symmetry is broken. To arrive at a unified theory of electro-weak interactions, physicists had to take into account a spontaneous symmetry break of this kind.

Some of today's famous theorists have gone so far as to claim that modern theories of physics can be guided purely by symmetry and nature's beauty. These scientists should keep in mind that Pauli's intuition was guided by symmetry and that he was wrong several times, most notably about parity violation. Yet while Pauli sometimes scoffed at theories that turned out to be correct, he never let an incorrect theory get by him.

The real power of modern science is in the synergy between experimental and theoretical physics. As we've seen, theoretical physics helps make sense of experimental results, and experimental results either reinforce or call into question the correctness of theory. Remember that even Pauli, when confronted with experimental proof, quickly withdrew any objections.

Although some professional physicists may quarrel with the way I have distributed scientific credit in this chapter, I call to my defense Rutherford. When Chadwick, working in Rutherford's lab, announced the discovery of the neutron, he and Rutherford received a nasty letter from a theorist whom Chadwick and Rutherford had chosen not to credit for his theoretical speculation a year earlier about just such a possibility. Rutherford claimed that he himself had speculated on the existence of the neutron in public lectures several years previously, and that speculation about such a particle among the members of his group went back over five years. He said it best when he replied to this physicist, "It is very easy to write about those matters, but more difficult to get experimental evidence." The debate between experimentalist and theorist over which gets most of the credit will surely continue. In defense of theorists, however, keep in mind that this author is an experimentalist and not unbiased.

The scientific community has given much thought to the theoretical formulation of electro-weak interactions since the discovery of neutral-current neutrino interactions. The theory has been greatly refined, and theorists are now able to predict with greater accuracy any quantity that experimenters want to measure. For the development of this theory,

Salam, Glashow, and Weinberg were awarded the coveted Nobel Prize in physics in 1979.

I have paid little attention so far in this book to who received the Nobel Prize for the discoveries it describes. At this time, however, I would like to raise a few points about the prize that may provide another lesson in the scientific process.

The Nobel Prize, awarded annually by the Swedish Academy of Science, goes to individuals and has no strings attached; the recipients do with the prize money what they please. The prize is never awarded to deceased scientists. If it were, the prize committee would have probably by now given the prize to Newton or Galileo. In the case of the neutral-current neutrino interaction, the founder of the experiment leading to the discovery of these interactions, Lagarrigue, and the second most important scientist on that project, Musset, would have been logical choices for the prize, but both men died unexpectedly and could not receive the award. In the reality of the scientific world, however, most scientists do not need the endorsement of a Nobel Prize to receive recognition for great achievements. Instead, the Nobel Prize provides a very good forum for disseminating scientific achievement to the front page of the newspaper. For its recipients, the prize provides recognition outside their scientific circles.

Winning the Nobel Prize involves a little bit of luck. Being at the right place at the right time has been important in some cases. The development of new equipment or new techniques may have made a particular discovery easier. The shear quantity of competition is so great that many bright, creative scientists will never get the prize. The Nobel Prize committee has even on occasion overlooked important people, sometimes by accident and sometimes because a scientist's work did not receive adequate publicity.

The recognition bestowed on a Nobel Prize–winning scientist by the public, governments, and the press has sometimes been useful. Einstein's stature as a Nobel Prize winner can explain in part the degree to which his letter to President Roosevelt, urging him to initiate a project to build a nuclear bomb, influenced Roosevelt's decision to commission the Manhattan Project. Sometimes, however, the recognition is not a healthy influence on the scientists themselves. Unfortunately, either ambition to win the prize or the extraordinary attention bestowed upon prize winners has altered the better nature of some good physicists for the worse.

In the past, the prize was customarily awarded after the accumulation of many other smaller prizes. It was a means, then, of providing the most inventive scientists with a monetary reward for their ingenuity. Today, however, many of these smaller prizes go to the person who has just received the Nobel Prize, most likely because the smaller awards committees either don't have the means to learn about other important scientific results or because they don't wish to be left out in honoring someone they had previously overlooked.

The Nobel Prize is usually awarded long after the work for which it has been bestowed is accomplished, because time is needed either to duplicate results or to put the work into its proper historical perspective. The significance of a discovery often isn't realized until other important work follows on its heels. So, if you ever do anything great in science and you want the Nobel Prize, try to live a healthy life so that your achievements will eventually be recognized!

From Theories on Paper to Computer Modeling

Once the electro-weak theory had been formulated, it turned out to be quite complicated, and its equations were not easily solved by traditional means. Fortunately newer methods were available. A different kind of calculation technique based on random numbers had come into use in the 1940s, and its employment would blossom with the invention of the computer. By the time the electro-weak theory had been formulated, faster and more powerful computers had transformed this new computational mathematics into a standard calculation technique of particle physics.

The new calculation technique, called the Monte Carlo technique, had its origins in the numerical calculations that physicists had to perform for the Manhattan Project. The procedure can, in effect, be played out at a casino in Las Vegas or Monte Carlo, which is where it obviously gets its name. "Monte-Carlo" is a term heard over and over again in discussions about particle physics. The best description of this mathematical technique comes from its use in its earliest days.

In classical physics, a single equation or a group of equations are sufficient to describe physical phenomena. The same is also true for fundamental quantum mechanics and elementary particle interactions. This relatively straightforward situation became more complicated with the introduction of the modern field-theory formulation of particle interactions after World War II. In the quantum world, how a particle propagates and whether it interacts are based upon probabilities; as these depend on many variables, it is difficult to predict what a particle will do. For instance, imagine a neutron in a nuclear reactor. We know that when this neutron is slowed down in the reactor's heavy water, it can interact with a nucleus and break it apart, as a cue ball breaks up the rack in billiards. However, the mathematical description of this interaction is a bit difficult, requiring a massive number of calculations. So instead of deriving one formula, the physicist creates a simulation of the event that takes into account the probabilities of the neutron performing certain actions. The neutron could travel through space without interacting, or it could be captured by a nucleus, or it could break up a nucleus to form fission fragments of the nucleus and more neutrons.

For each action that the neutron can take, it is assigned a number on a roulette wheel. To play the "game" of Monte Carlo, the physicist considers a collection of many neutrons. To discover their statistical fate, he or she spins the wheel once for each step a neutron takes and then determines what has happened to the neutrons. Eventually, a single neutron interacts with a nucleus to produce more neutrons, and their fate must also be determined statistically by spinning the wheel. The process either terminates in the complete elimination of all the neutrons, or it escalates the number of neutrons in such tremendous proportions that a self-sustaining nuclear reaction results. Physicists performing these calculations can then begin the game over again, but this time they would use completely different parameters—they could change the neutron's initial energy, the nuclei isotopes in the reactor, or the moderator material (heavy water, normal water, or graphite). By altering the parameters they can determine the best way to construct the reaction in order to produce the intended result.

In the early days, the calculations were actually performed with Vegas-style roulette wheels. Although physicists no longer use roulette wheels, the basic components of the Monte Carlo calculation technique are as dependent on the probabilistic nature of quantum mechanics as gamblers (and casino owners) are on the probabilistic nature of the wheel.

Today, the experimental apparatus as well as the detailed equations of particle physics theories are so complicated that determining how one parameter affects the entire process might be impossible from the theory itself. It is in such cases that a physicist employs numerical calculations using the Monte Carlo technique. Instead of roulette wheels, the physicist uses a computer equipped with a random number generator. Because these generators never repeat a sequence of numbers, they are as unpredictable as a roulette wheel. A numerical simulation is done over and over again, providing results that should reflect what actually occurs when experimental measurements are taken. If the measured results do not match the Monte Carlo calculations, the error usually lies somewhere in the setting of the parameters or in the initial complicated equation itself. If the parameters cannot be adjusted to match the observed results, then the theory itself is viewed as not being a complete representation of nature. However, in cases where the adjusted parameters produce matching results, Monte Carlo is a useful tool for predicting what will happen in similar situations.

The fact that statistics and probability play such a big role in the calculations of modern physics models does not mean that the models are less accurate than those of classical physics. Rather, it means that these models cannot be represented by one easily solved equation on a single line of paper. To truly understand such a model, the physicist must encode the correct computations into the computer and have the computer explicitly search for dependencies between parameters and interactions. While it is true that a physicist can more easily see the basic relationships of a model if it is composed of one or a group of equations, the more complicated models have one advantage. Once a model is fully encoded into a computer program, the physicist can enter any conditions into the program and quickly obtain a result. The fact that the result is not an equation but is similar to the measurements taken in experiments is only a problem for especially puritanical theorists.

Many of the results discussed in subsequent chapters are based on numerical and Monte Carlo calculations. Keep these methods in mind when reading about the models of weak interactions, supernova explosions, and nuclei synthesis in Chapters 8 and 9.

The Proton-Antiproton Collider

The current unified electro-weak theory, refined by the contributions of many scientists, provides an explanation for the properties of neutral-current interactions. At this point in our story, however, no experiment had confirmed the existence of the most crucial pieces of the theory, the particles that propagate the weak nuclear force—the positive and negative W particles and the neutral Z^0 particle. Their existence was almost certain, and it was possible to estimate their masses, but at the time of the Gargamelle experiment and the formulation of the first unified theory of electro-weak interactions, these particles had not yet been created for direct study in the laboratory. There was some doubt about the actual existence of the neutral Z^0 particle; some physicists thought it might just be a figment of the theorists' mathematics. Very little was known about the Z^0 particle's interactions other than neutral-current neutrino interactions, and the possibility always existed that what seemed to be the interaction of a Z^0 particle might really be a neutral interaction involving both a positive and negative W particle simultaneously. Such possibilities must be kept in mind until a theoretical speculation is verified.

How much longer scientists would have to wait before finding out whether the W and Z^0 particles actually exist became a pressing question. To prove their existence, these particles would have to be created in real form under laboratory conditions. A real particle, unlike a virtual particle exchanged during an interaction, does more than just carry a force; it also has a well-defined mass, and it propagates through time with a well-defined lifetime. Before anyone could claim that the weak-force propagator particle had been observed, many W particles would have to be produced; these particles would permit a measurement of mass as well as a study of any other interactions they might participate in and properties they might possess. The fact that Salam, Glashow, and Weinberg could deduce the theoretical properties of the electro-weak force without actually producing the particle that propagates this force was a triumph truly deserving of the Nobel Prize. These great theoretical achievements would soon be verified in a forthcoming experiment.

The Gargamelle experiment had obtained its neutrino beam from the small Proton Synchrotron (PS) accelerator, which could accelerate particles only to 26 GeV, a relatively low energy. Yet even before the experimenters had analyzed their many pictures of neutrino interaction events obtained from this accelerator, CERN was preparing the designs for a larger, higher-energy accelerator that would permit the production of real W and Z^0 particles. This new accelerator was going to be on the scale of the accelerator just completed at Fermilab, but would have a slightly higher energy. CERN had one great advantage over Fermilab: it was able to expand its complex so that all that had to be built was the new accelerator, whereas Fermilab's entire accelerator complex and infrastructure had to be built from scratch. CERN's existing accelerators were used to provide a fully functional low-energy accelerated beam to put into the new accelerator, saving time, effort, and, most important, money. CERN also had an existing lab staffed with talented teams of scientists; again, Fermilab's teams had yet to mature scientifically by the time of the Gargamelle discovery. CERN's new Super Proton Synchrotron (SPS) was constructed in the first half of the 1970s and turned on during the summer of 1976.

Although the many experiments in the CERN-SPS and Fermilab fixed-target experimental programs performed some very good physics, none reached an energy sufficient to produce the weak-force propagator particle. To finally achieve this important goal, CERN's accelerator physicist, Simon Van der Meer, came up with an inventive redesign of the SPS. Instead of accelerating protons to an energy of 320 GeV, at which point they were transported to an experiment's fixed target, Van der Meer and his team created a technique that produced a large number of antiprotons. These were accumulated into a low-energy beam, then injected into the large SPS accelerator at the same time as the protons, but aimed in the opposite direction. In magnetic fields designed for protons, antiprotons travel in the opposite direction to their normal-matter counterparts. CERN could thus create two beams of particles, each at the high energy of 270 GeV, that were counterrotating in the same accelerator. These two beams were carefully controlled to prevent them from colliding at any but two locations in the ring. These were the locations at which experiments could be built. The head-on collisions between the proton and antiproton beams, collisions that annihilate both particles, produce much more energy than a fixed-target experiment could, and this energy was harnessed to produce particles heavier

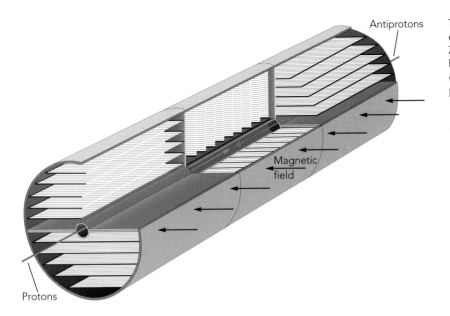

Antiprotons

Magnetic field

Protons

The inner tracking chamber of the UA1 experiment, which first observed the W and Z^0 particles. This detector is a modern type having thousands of fine wires in a gas-filled chamber; it permits electronic readout of particle interactions at a higher rate.

than had been possible before. The redesigned CERN accelerator complex was complete and ready for experimentation by 1981.

In the decade between the Gargamelle bubble chamber experiment and the completion of the first proton-antiproton collider, the technology for measuring particles had drastically changed. Physicists using a bubble chamber could measure only the curvature of a particle in a magnetic field as revealed by a trail of bubbles in the liquid. They could also see decays of particles in flight. Each type of particle generated a signature trail of bubbles by which it could be identified only poorly. Furthermore, although a bubble chamber was large, it was a relatively simple device. It contained one large volume of liquid that was viewed with many stereoscopic cameras. It was a basic, one-unit piece of apparatus.

Ten years later, the bubble chamber was gone, and a detector built of many different components had taken its place. In 1969 Georges Charpak had invented a new kind of electronic detector called a multi-wire proportional chamber. This type of detector comes in many different varieties, but all of them have a gas-filled volume crisscrossed by hundreds if not thousands of fine wires that do the tracking. These gas-filled detectors were able to collect data at a higher rate than bubble chambers; this ability, as we saw earlier in relation to the Gargamelle data, is especially useful when searching for rare processes or special events. Physicists also began routinely measuring energy in calorimeters to obtain more information about the particles coming out of any interaction. However, even though the modern electronic detector of the

The UA1 detector being prepared for operation. The detector, the size of a three-story office building, has proton-antiproton collisions occurring at its center. The collision point is surrounded by the gas-filled wire chamber shown in the previous figure, and that is in turn surrounded by an energy-measuring device. The signals from the detector travel along the many round cables seen coming out at the bottom toward the camera.

1980s and '90s is very different from the bubble chambers of the 1970s, there are many similarities in the analysis of the particles. Both detectors can measure only charged particles, since neutral particles, no matter what species, cannot leave a track behind to analyze. Both can measure the energy and momentum of a particle. And finally, the design of any kind of detector experiment determines what can be done with the data after they are collected.

CERN researchers built detectors for separate experiments, called UA1 and UA2, at the two points in the SPS collider where the proton-antiproton particle beams intersected. The "UA" is an abbreviation for "underground area," and reflects the fact that both experiments were performed underground in order to prevent particles produced in the collision from leaking radiation onto the Earth's surface. The most important of these experiments, led by Carlo Rubbia, was UA1, since its detector was superior in momentum resolution and performance to UA2's. Rubbia was also the major instigator behind the new collider. Had he not motivated CERN and the accelerator physicists to trans-

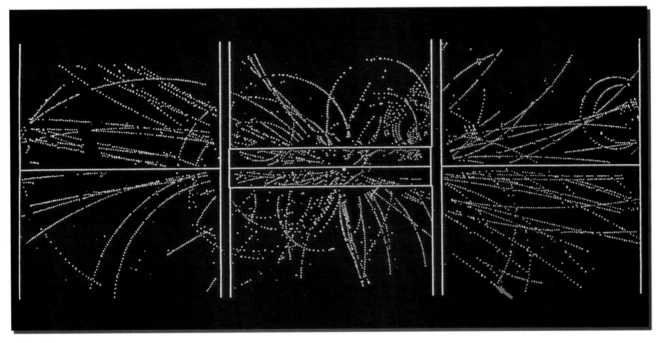

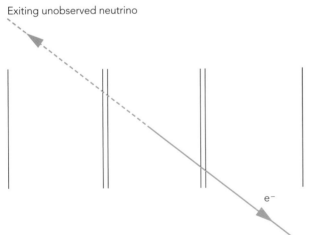

Exiting unobserved neutrino

e⁻

A W particle event seen by the electronic detectors of the UA1 experiment. The red arrow at the bottom right points to the track of the particle carrying most of the energy, identified as an electron. The sketch shows this track as well as the calculated track of an unobserved neutrino produced in the decay of the W.

form the accelerator into a collider, the experiment would not have been completed in as timely a fashion, and the ultimate detection of the W and Z^0 particles would have been delayed considerably.

When a particle from the antiproton beam collides with a proton, an antiquark from the antiproton collides with a quark from the proton. The collision produces an energetic photon as the two quarks are

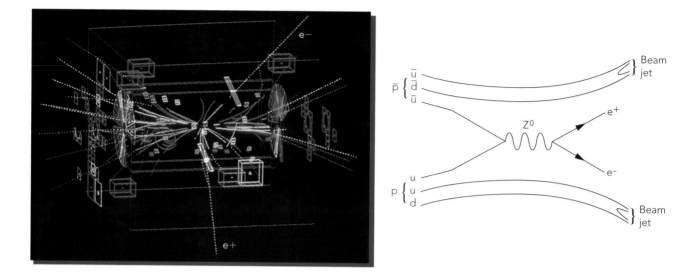

The collision of a proton and an antiproton, producing a Z⁰ particle, as seen in the UA1 detector. The Z⁰ was observed by seeing its decay into an electron-positron pair going off in opposite directions, (blue tracks). The remaining tracks are hadronic matter such as pions and kaons, formed by the fragments of the proton and antiproton that did not react to produce the Z⁰.

annihilated. The other quarks in the proton and antiproton do not annihilate and are responsible for the large number of low-energy particles produced in these very messy interactions. When the collision is at sufficient energy, it produces real weak-force propagator particles, the W and Z⁰. These events have very special properties that make them easy to find. The large amount of energy contained in one electron track, produced from the W's decay, is the signature for that particle's production. The Z⁰'s signature would be a pair of high-energy electrons or muons of opposite charge going out in opposite directions. Both kinds of events were found in large numbers in the data, and physicists were able to measure the mass of the particles as 80 GeV/c^2 for the W and 91 GeV/c^2 for the Z⁰. These mass measurements became valuable input for improving the electro-weak theory, as did measurements that allowed physicists to understand the precise mixing of the strange and down quarks, and to make detailed studies of the decay of Z⁰ and W into normal matter.

The mass measurements were important for another reason. The W and the Z⁰ were the first force-propagator particles found to have mass—the photon of electromagnetism and the gluon of the strong nuclear force are both massless. As explained in Chapter 3, it is because the weak-force propagator particle actually has mass that it acts over such a short range and has such a small probability of interaction. The mass measurements were a final proof of this concept.

By 1983, the long search for the charged mediator particle first predicted by Yukawa was over. In the decades since Yukawa proposed his theory, much has been learned about the weak force, including the existence of another weak-force mediator particle, this one neutral. In the end, the theory looked very different from Yukawa's initial proposal, but his input back in the early 1930s was clearly necessary to help get physicists thinking and searching in the right direction.

. . .

Although Yukawa's theory of weak interactions, with its postulated mediator particle, was aesthetically pleasing on paper, it still seemed to be missing something because it did not very accurately predict experimental results. It took the discovery of a new neutral weak-force mediator particle, the Z^0, to point theorists confidently toward the weak interaction theory that we know today. This discovery was CERN's first great experimental triumph, but its success grew out of its failed first neutrino experiments.

Physicists established the indirect existence of the Z^0 particle by studying neutrino interactions in a unique heavy-liquid bubble chamber. Their results led to the vindication of the unification of the electromagnetic force with the weak nuclear force into one complete theory that could describe both. Not since the time of Maxwell in the nineteenth century had two forces been merged into one mathematical formulation, but the incredible feat of merging weak-nuclear interactions into the electromagnetic force was an even more thrilling feat because physicists had devised the complete theory without producing any real weak-force mediator particles, the W and Z^0, for actual study.

To investigate the question of whether the W and Z^0 particles actually existed as observable particles, laboratories around the world began to construct new accelerators. I have described only a few of these laboratories and their accelerators in this chapter. The United States built at least four machines and Japan one, while in Europe, at CERN, the proton-antiproton collider that actually produced the first observable W and Z^0 particles was in fact the fourth machine in a series of accelerators that had helped in the search. Although the discovery did not come until 1983, the design of the SPS had begun in the late 1960s, and the machine had evolved into the proton-antiproton collider by the late 1970s through the persistent efforts of physicists determined to confirm the theory of electro-weak interactions. Many theories were discarded and many surprises uncovered in the course of the experiments that guided the way to the final and correct theory.

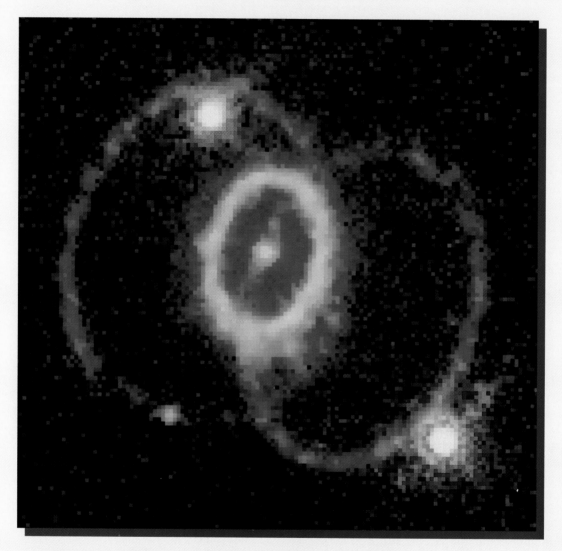

These three rings and a diffuse core, shown in an image produced in 1994 by the Hubble Space Telescope, are the remnants of a supernova explosion detected on Earth in 1987. The blast sent into space a surge of neutrinos that triggered underground detectors on both sides of the globe.

CHAPTER 8

Exploring the Universe with a Neutrino Telescope

Astronomers discovered new wonders when they began to look at the universe beyond the visible spectrum. The night sky appeared very different when, instead of mapping the light emitted by stellar objects, they mapped the invisible radio waves, X-rays, and gamma rays from the lower and upper ends of the electromagnetic spectrum. Among the most interesting objects they discovered were pulsar stars, galactic cores bilging jets of matter, and possible candidates for black holes.

Although neutrinos and the photons of light are both particles, any similarity between them stops there. Neutrinos are not part of the electromagnetic spectrum commonly used to explore the night sky, but they can serve a similar purpose as a window on the universe. The first hint of their usefulness came in 1987, when physicists detected a neutrino burst from an exploding supernova. Although they had expected neutrinos to be released in a supernova explosion, the pattern of their emission has compelled astronomers to rethink the theory of these explosions. Specially constructed neutrino detectors might provide us with a means to search the sky for further supernovas and other neutrino-emitting objects, and allow us to study these objects in greater detail than previously possible.

Looking into the Unknown

The idea that an instrument built to detect particles such as neutrinos could be a kind of telescope may seem odd. Most people think of a telescope as a set of lenses that focus light from distant objects, causing the images of these objects to become magnified. But throughout the last hundred years the definition of a telescope has been broadening to include instruments that detect other forms of radiation, such as X-rays and gamma rays. Those instruments may operate on principles quite different from the focusing of electromagnetic waves employed by an optical telescope. The main defining property of a telescope then becomes not the type of emission detected, or the focusing of it, but the ability to pinpoint the source of the emission in the sky. Hence the proper definition of a telescope is any instrument that looks for any form of radiation emitted by an object in order to image the emitting source.

Scientists spend much of their time studying unexplained but observable phenomena; sometimes they have to look for the as-yet-unobserved in order to expand our understanding of the universe. Seeking evidence for new phenomena is extremely difficult: Where does one begin and what, exactly, does one look for? Astronomers have been searching the sky for new phenomena for several hundred years, exploring the heavens with optical telescopes. In the centuries since its invention, the telescope has repeatedly been improved or modified to suit a particular purpose. Most recently, we have seen the launch of the Hubble Space Telescope, a relatively small, ordinary telescope that travels to uncommon places. The Hubble telescope has the advantage of being in orbit high above the Earth's atmosphere, from where it provides a new, clear window into the universe. It is essentially just a larger version of Galileo's telescope, and its purpose is the same—to seek and explore mysterious objects in the night sky.

Although optical images of stars and other objects provide exhilarating and inspiring images, the optical part of the photon spectrum

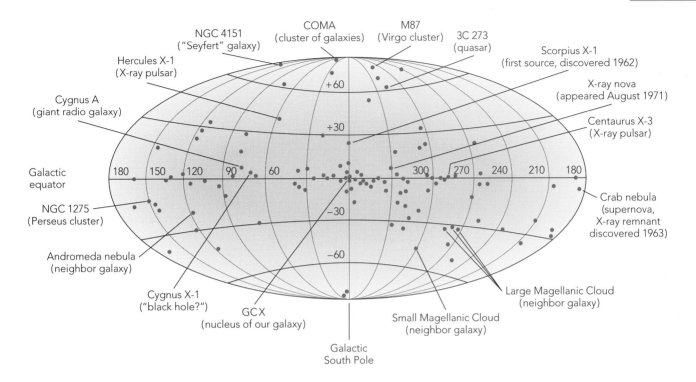

NGC 4151
("Seyfert" galaxy)

COMA
(cluster of galaxies)

M87
(Virgo cluster)

3C 273
(quasar)

Scorpius X-1
(first source, discovered 1962)

Hercules X-1
(X-ray pulsar)

X-ray nova
(appeared August 1971)

Cygnus A
(giant radio galaxy)

+60

+30

Centaurus X-3
(X-ray pulsar)

Galactic
equator

180   150   120   90   60   300   270   240   210   180

Crab nebula
(supernova,
X-ray remnant
discovered 1963)

NGC 1275
(Perseus cluster)

−30

Andromeda nebula
(neighbor galaxy)

−60

Cygnus X-1
("black hole?")

GC X
(nucleus of our galaxy)

Small Magellanic Cloud
(neighbor galaxy)

Large Magellanic Cloud
(neighbor galaxy)

Galactic
South Pole

A sky map of X-ray-emitting sources seen by the first X-ray satellite observatory, UHURU.

represents only a small fraction of the energy output of stars and galaxies. The night sky viewed in infrared light, X-ray spectra, or radio signals looks distinctly different from the sky seen with an optical telescope. When astronomers began to look beyond the visible spectrum at ultraviolet or infrared light, or at X-rays and high-energy gamma rays and radio waves, the information they gathered helped provide an explanation of the dynamics and origins of objects that had seemed mysterious when viewed solely by visible light.

For example, X-ray–detecting satellites launched in orbit high above Earth's atmosphere helped lead the search for exotic objects such as black holes—high concentrations of matter that have collapsed into an infinitesimally small point called a singularity. The gravitational attraction of a black hole is so great that no known particle is energetic enough to escape from one, not even light. A black hole has no direct optical emissions and can only be "seen" by observing phenomena just outside the hole's boundary. Black hole theories, which originated in the fields of relativity and cosmology, predicted that these objects would be intense emitters of X-rays. This type of radiation would be given off when matter fell through the strong gravitational fields near the limit of the black hole's event horizon, a spherical limit around the hole marking the boundary beyond which nothing can escape. The satellites searching for these X-rays found only a few intense X-ray emitters,

An X-ray view of the center of our galaxy. The many bright round patches are point emitters of X-rays most likely produced by matter falling into neutron stars. The orange bands represent a diffuse emission of X-rays possibly given off by matter falling into a massive black hole at the center of the galaxy. The dark band through the middle and in the top center indicates the presence of galactic dust clouds that X-rays cannot penetrate.

along with a diffuse background of X-rays in the plane of our galaxy. At first, astronomers considered all X-ray–emitting objects to be candidates for black holes, but after studying the objects carefully they realized that not all were alike. Some turned out to be diffuse sources of X-ray emissions; others emitted X-rays from a single, highly concentrated point. After further detailed imaging by X-ray satellites, only the pointlike X-ray emitters are still considered good black hole candidates.

Our Sun is also a source of X-ray emission; its X-rays come from charged particles emitted by giant solar flares. These flares emerge from the sun in huge arches that emit X-rays as they spiral in the intense magnetic fields near the Sun's surface. The flares eject hot gases, called plasmas, composed of charged particles that may travel all the way to Earth, where they are trapped by our planet's magnetic field and funneled into the polar regions. There these particles cause the spectacular aurora borealis, or Swedish lights, that can be seen from northern countries such as Canada or Sweden. By observing the Sun in the X-ray spectrum, scientists see slightly deeper into the solar layers than they can see with optical telescopes. However, as described in Chapter 4, only the neutrino is capable of giving a view into the Sun's central core.

Astronomers have also studied X-ray emissions from several galactic cores. When they observe only the lower-energy "soft" X-rays, from 5 to 20 keV, the cores appear to be large, diffuse objects. But when they

Right: A cutaway view of EGRET (Energetic Gamma-Ray Experimental Telescope), an instrument carried by the Gamma-Ray Observatory satellite. At the top is a small spark chamber from which data is read out electronically. Incoming gamma rays interact with a material of high mass that sits between each of the spark chamber's many layers, producing electron-positron pairs detectable by the spark chamber. Underneath the spark chamber is a device that provides a measure of the gamma ray's initial energy. Left: A view of our universe as seen through the eyes of EGRET reveals highly energetic gamma rays in the galactic plane.

observe X-rays at energies greater than 100 keV, the X-ray sources appear to be single pointlike emitters within the diffuse clouds. The diffusion of the soft X-rays is created by the high concentration of matter in the cores of galaxies: high-energy X-rays scatter off the matter, becoming diffuse, and lose energy, becoming softer. The Andromeda galaxy—the nearest spiral galaxy to our own—has two core emission regions; this observation has incited a great deal of excitement, and the origin of the twin cores a great deal of speculation. Yet their origin remains unclear even though the Hubble telescope has been able to finally observe these two regions of the Andromeda galaxy. The study of other spectral emissions from these and other mysterious sources may yet be revealing.

Even more energetic than X-rays are gamma rays. In 1972 NASA provided a spectacular way for particle physicists to observe the gamma-ray sky when it launched its SAS-2 satellite—a particle physics observatory in space. NASA had equipped the SAS-2 observatory with a spark chamber similar to but smaller than the one described in Chapter 5. The satellite's spark chamber stack recorded the direction from which a charged particle in the chamber had traveled so that astronomers and physicists could identify the source of the particles.

Charged particles themselves cannot be used to detect emission sources because the magnetic fields generated by our Earth, Sun, and galaxy bend the particles, leaving no trace of their initial origin. However, photons that make it to Earth from distant objects are unaffected by any magnetic fields. In the spark chamber, a high-energy photon is converted into an electron-positron pair; a simple computation reconstructs the initial photon's direction of travel from the paths that the pair follow in the chamber. By collecting such converted gamma-ray events over nearly a year, the SAS-2 satellite was able to map the high-energy gamma-ray sky. An improved version of this gamma ray telescope, COS-B, operated for nearly seven years, providing an even better map.

Many of the gamma-ray emitters recorded in this manner were linked with X-ray and optical sources. In one case, the most recent X-ray satellite, dubbed the Einstein Observatory, measured structures within the brightest X-ray objects, finding strong evidence that one of these objects shielded a black hole. Astronomers have firmed their identification of a black hole through correlating these X-ray emissions with gamma-ray observations and optical observations made with both ground-based telescopes and the Hubble telescope.

A true black hole is distinctly different from any other object in at least two respects: it can be extremely massive and it lacks a surface that can be directly observed. As we've seen, initial X-ray observations of the sky pointed to many objects that were intense X-ray emitters. Before 1992, the best black hole candidate was considered to be an X-ray emitter that had been given the name Cygnus X-1. However, Cygnus X-1's mass may have been drastically overestimated; since the object itself cannot be seen, the mass had to be inferred by observing the rotation of a companion star. Today, an object called V404 in the Cygni constellation is considered the best black hole candidate; indeed, many astrophysicists believe that no other phenomenon could explain it. V404 has an intense X-ray emission of the type that should be present when matter falls into the sort of gravitational accretion disk that is believed to surround black holes. Its lack of any gamma-ray signal has also helped to confirm V404 as a candidate. Gamma-ray emissions are not expected, since black holes lack any solid surface visible to an observer outside the hole, and gamma rays are usually produced only when matter strikes other forms of matter such as the surface of a neutron star or white dwarf star. The conclusive piece of evidence is the extraordinarily fast rotation of a companion star in orbit around V404. Even if the gas composing the companion star were of featherweight density—as is not possible since a star must have a minimum density of mass in order to ignite—we could still infer that V404 had *twice* the amount of mass predicted for a black hole to form. Based on the evidence, the verdict is in: the unseen object can be nothing other than a black hole.

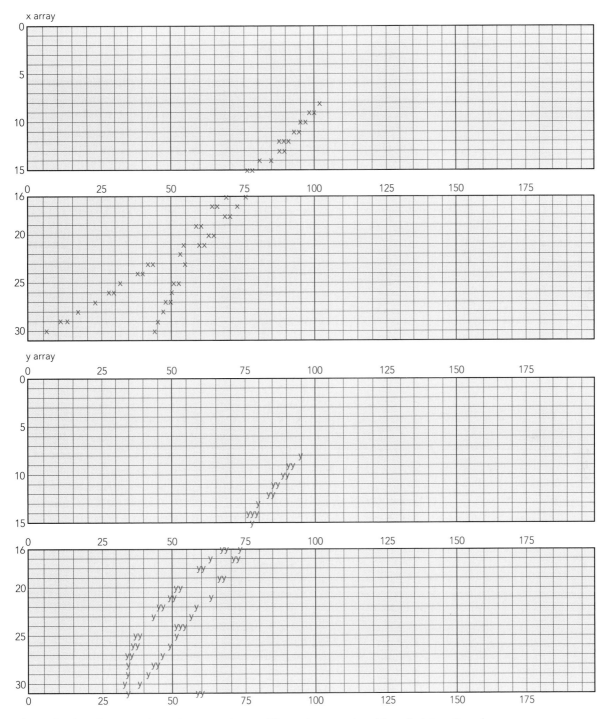

The tracks of an electron and a positron, seen in two different views, produced from the interaction of a gamma ray in the SAS-2 spark chamber. By measuring their direction of travel, scientists could precisely reconstruct the initial gamma-ray direction.

The Hubble telescope has recently revealed a disk of glowing gas orbiting the core of the giant elliptical galaxy M87. The disk has been attributed to matter falling into a massive black hole.

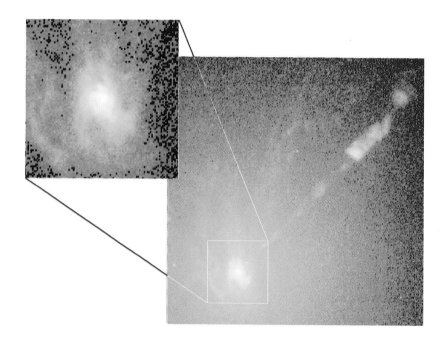

The galactic core of the elliptical galaxy M87, an intense emitter of both X-rays and gamma rays, was selected to be one of the first objects scrutinized by the improved Hubble Space Telescope. The Hubble telescope found matter spiraling around the core of M87 at a phenomenal velocity. The orbital velocity of the gas is so high that no phenomenon other than the accretion disk of a black hole can explain it. In this case, because the black hole is at the center of a galaxy, and so must have gargantuan mass, matter falling through its immense gravitational fields emits gamma rays as well as X-rays.

The Deaths of Stars

I have yet to offer an explanation for the source of the massive objects called black holes. If we trace the final stages of a large star's life, we will be led to the beginning stages of a black hole's life. We will find that a star's death brings us back to the story of the neutrino, for a dying star was the first object in distant space to be recorded on Earth by a neutrino detector.

As we saw in Chapter 4, a star continues to burn heavier elements as they accumulate in its core. After burning carbon, a star next burns nickel, in a reaction that produces iron. As this process continues, the core of the star becomes denser. If the original star had a low mass, then the density of matter in the core does not create a gravitational pressure great enough to cause the star to collapse; the repulsion between identically charged nuclei is capable of holding up the weight of the star against gravitational pressure. The star will instead burn out, leaving a rock in space. This kind of star is called a white dwarf. Its light slowly fades, but it remains stable, provided no more mass is added.

If the original star has sufficient mass, the immense gravitational pressure in the core cannot be withstood. The electrons are squeezed into the protons, and the star collapses into an object made almost entirely of neutrons. The matter composing a neutron star is so compact that, on average, such a star is only seven miles across.

No physical laws are violated in the process of a star's collapse. Electrons interact with protons to form neutrons, releasing neutrinos through the weak interaction, as shown in the diagram in the margin. A star's collapse into a neutron star is almost instantaneous, and the number of neutrinos produced is immense. As it collapses, the star's outer layers are blown away in a supernova explosion. Left surrounding the neutron star are the remnants of the supernova, a glowing gas cloud called a nebula.

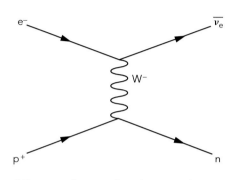

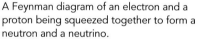

A Feynman diagram of an electron and a proton being squeezed together to form a neutron and a neutrino.

The nebula formed by a supernova explosion is one of the most beautiful objects in the night sky. The inspiring Crab nebula we see today can be traced back to the year 1006, when astrologers in China witnessed an immense new star, so bright that it could be seen in the daylight at noon for a month. This "new" star was really an old star undergoing supernova explosion. The Crab and other nebulae may owe their existence to the immense neutrino pressure that builds up inside a collapsing star. The neutrinos escaping the inner core encounter matter at such a high density that a large number of neutrinos interact; in doing so they heat the outer layers of the star, blowing them away. The only effect of the neutrino visible to the human eye might just be these wondrous nebulae.

If a massive star collapses to form a neutron star, what happens to a supermassive star? Such a star, while collapsing into a neutron star, could by the force of gravity collapse still further into a singularity—a black hole. But even if a star's collapse is first halted at the neutron star stage, its gravitational field is so strong that over time it could continue to accumulate matter. Eventually, after enough matter had accumulated, this neutron star and its additional matter could collapse into a black hole. Both these paths to the formation of a black hole occur because the star cannot support its weight against the gravitational pressure. The huge accumulated mass generates sufficient pressure to overcome the strong nuclear force holding the neutrons apart.

Black holes have received a great deal of attention from a public amazed that so much matter could be concentrated in one point. The question that continues to bewilder astronomers is what happens inside a black hole. Theoretical models tell us that the gravitational attraction

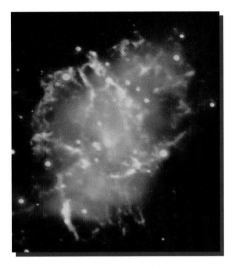

The Crab nebula, formed from the supernova of 1006, harbors a neutron star at its core.

expected by a black hole should be so great that it bends space around itself so that not even light or, presumably, neutrinos can get out. Since we cannot see into a black hole without traveling inside it, we do not know what type of structure exists at its core, and statements claiming that the laws of particle physics are violated during the collapse of a star are mere speculation, reflecting our own scientific ignorance. What we do know is that the laws of particle physics in such circumstances are different from anything we can study in the laboratory.

Recently, particle physicists have been considering states of matter intermediate between those of the neutron star and black hole. Neutrons are made only of up and down quarks, the lightest and simplest building blocks of nucleons. A free neutron is unstable and decays, but when a neutron is bound to a proton, the nuclear force binding the particles keeps it as stable as the proton. Although a neutron star may have no protons, its neutrons are stable because they are bound by gravity, which exerts a powerful pressure in such a massive object. We know from experimental particle physics that nucleons made of strange or charm quarks are also unstable, but it is possible that the immense gravitational pressure at the core of a neutron star can allow denser forms of matter to exist, perhaps even strange matter. These speculative quark stars would not be able to sustain their own weight if composed only of the up and down quarks of normal neutrons, but they may be sustainable if the quarks are those belonging to the next generation of matter—strange and charm quarks. The possibility exists that the quarks of heavier nucleons, such as the lambda, can also be stabilized by the gravitational potential found in a collapsed star, similar to the way in which neutrons are stabilized in a conventional neutron star. If this is true, it would be more difficult to form a black hole—collapsing stars would have to be more massive.

Neutrinos from the Supernova of 1987

On the twenty-third of February 1987, in the irregular galaxies close to our own galaxy in what is called the Small Magellanic Cloud, a star went supernova, the closest supernova to our solar system in 380 years. Astronomers first spotted the exploding star in photographic plates taken that night, while physicists running experiments conducted deep underground saw neutrinos that had arrived straight from the explosion. The appearance of the supernova would provide a splendid opportunity to test modern theories of stellar collapse. When all the scientific analysis was done, scientists would have changed the way they think about supernova explosions.

Several large underground experiments sensitive to neutrinos were operational during this supernova explosion. These experiments had been originally designed to search for proton decay, as I will discuss in

Chapter 10, but the detectors used in these experiments were also able to function as the first neutrino telescopes. Since the Magellanic Clouds are not in our galaxy, these detectors were truly the first neutrino telescopes used to explore the universe.

Two of these detectors, one located in Cleveland, Ohio, and the other in Kamiokande, Japan, saw neutrino signals that correlated with the supernova. Both detectors are large tanks of water lined with photodetectors to observe Cherenkov light. They are situated deep underground; the detector in Cleveland had been placed in the Morton salt mines underneath Lake Erie. Both detectors found neutrinos through a process different from those described in Chapter 4.

Neutrinos from a supernova explosion are higher in energy than the neutrinos emitted from a reactor or from the core of the Sun. Consequently, the particles produced when the neutrinos interact within the detector also have a correspondingly higher energy. The most energetic

Physicist/scuba-diver Karl Luttrell repairs components of the photosensitive array inside the IMB Cherenkov detector. This photo, taken through 60 feet of water, shows the extreme clarity of the purified water essential to the propagation of Cherenkov light. This was one of the two detectors that saw the light pulses of converted neutrinos from the 1987 supernova explosion.

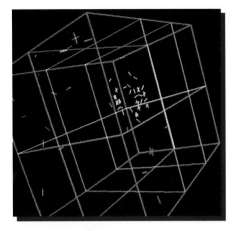

The arrival of a neutrino from the supernova of 1987, as seen by the IMB detector. The square box represents the walls of the detector, each one a different color, and the short yellow lines forming a ring represent Cherenkov light counts indicating the presence of an energetic lepton produced by a neutrino interaction.

particle produced is a charged lepton. This lepton can be tracked in a multiwire proportional chamber, like those described near the end of Chapter 7, or it can be observed optically. A highly energetic lepton of 20 MeV will emit Cherenkov light in a forward cone that photosensitive detectors can pick up. Such a detector reconstructs the incoming neutrino's direction of travel rather poorly, because it observes only one of the particles produced by the neutrino interaction. Nevertheless, it is able to crudely distinguish the direction from which an incoming neutrino has come to an accuracy of about three moon diameters. This is not good enough to create an image of a neutrino source's structure.

After optical astronomers had announced the supernova to the world, the scientists at the two large detectors searched their data for an unusual surge in neutrinos. They found a significant number of neutrinos that had arrived from the supernova, revealing a neutrino pulse with a puzzling structure. The Kamiokande researchers noted 11 high-energy neutrino events observed above a background of lower-energy events attributed to other origins; these 11 neutrinos arrived in two pulses separated by about 10 seconds. The Cleveland detector (called the IMB detector for the universities and research institutes that first created the experiment, the University of California at Irvine, the University of Michigan, and Brookhaven National Laboratory) had registered 8 neutrino events with a slightly different structure. Scientists could think of three possible origins for the two pulses in the Kamiokande data.

First, because detecting neutrinos is so difficult, the apparent delay in the arrival of the second pulse could have been the result of an inefficiency in the detection system itself. With such a low number of observed events, statistical fluctuations could make it appear as if there were two pulses.

Second, a difference in mass or path length could explain the arrival of the neutrinos in two separate bunches. The collapsing star must have produced a tremendous number of neutrinos for a neutrino signal of this strength to be seen on Earth. The strength of the neutrino signal unambiguously indicates that a neutron star was formed in this supernova explosion. A neutron star with a high-mass core of strange or charm quarks would have produced muon neutrinos, in addition to the common electron-type neutrinos of beta decay produced when neutrons form. The two types of neutrinos would have different mass *if* the neutrino were not massless; the slightly more massive neutrinos would not travel as fast as the lighter ones and would arrive at the detectors slightly later. Alternatively, since neutron stars are so massive, they will distort the space around them, perhaps causing some neutrinos to take slightly different paths to the Earth. In this way, a small time delay could possibly result.

Third, the star may have bounced when it collapsed. As the initial star collapses into a neutron star, a small fraction of the escaping neutrinos heat up the outer layers as they interact while trying to get out of the core, causing those layers to be blown off. The material in these layers then falls back into the neutron star, say, approximately 10 seconds later. The immense gravitational pressure of a neutron star converts the returning matter to neutrons, and in the process the star would emit the second neutrino signal that was observed. If the mass added by the second collapse is greater than the highest mass a neutron star can support, then the core of the star might be transformed into something even more exotic than neutrons, or it might become a black hole. Unfortunately, this supernova explosion was too far away for astronomers to determine whether a black hole was formed. We should be thankful, however, that the explosion did not occur very close to us—the number of neutrinos would have been much greater, but the shock wave of a very close-by supernova could have torn the Earth's fragile atmosphere apart.

A bouncing collapse is considered the scenario most likely to have produced the unexpected delay *if* these neutrino observations are truly two separate neutrino pulses. The need to account for a bouncing collapse has forced astronomers to modify their models for supernova explosions.

Even if neutrinos are massless, in principle they should still obey the laws of gravity as formulated in Einstein's theory of general relativity. Einstein's theory holds true over photons, which are massless, because matter bends and curves space. This is the reason that not even photons are capable of getting out of a black hole. Because neutrinos interact only through the weak nuclear force, scientists debated whether they obey the laws of general relativity. Some speculated that if the weak nuclear force does not "notice" the warping of space, a neutrino could travel straight out of a black hole. The supernova of 1987 provided an answer. During the supernova, the neutrino burst observed on Earth arrived within eight hours of the optical light pulse. Although the time of arrival of the neutrino burst is known to the exact second, the arrival time of the light pulse is not, because the first photographic plate showing the explosion was taken eight hours after the previous plate showing that region in the sky. However, we know that, because light is affected by the warping of space, even the light pulse does not take a perfectly straight path to Earth. Since the Magellanic Clouds are so far away (100,000 light-years), the light's travel time could easily have been a month longer than it would have been if no matter existed anywhere along its path. The fact that the neutrino burst did not arrive weeks earlier tells us that even neutrinos obey the same laws of curved spacetime as photons do.

After the supernova of 1987 and the spectacular observation of neutrinos from the collapse of a star, scientists at many neutrino detectors reviewed their old data, searching for any evidence of similar neutrino

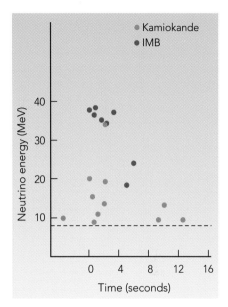

The energy and time of arrival of the supernova neutrinos from both the IMB and Kamiokande detectors. The dashed line indicates the sensitivity threshold of the detector.

bursts that might signal supernovae that optical telescopes had missed. Perhaps, for example, a star positioned behind a dense gas cloud had not been visible to a telescope from Earth. Six years' worth of data revealed no other neutrino bursts, but today newer experiments, to be described in the next sections, continue to keep vigil for signs of supernovae explosions.

The supernova explosion of 1987 gave physicists another estimate of the neutrino's mass: if that mass were extremely large, the particle would not have been able to travel anywhere near the speed of light. For a low-mass neutrino such as the electron-type neutrino produced in the supernova explosion, the spread in energy and the difference in arrival times at Earth put a limit on the mass at less than 10 MeV/c^2. This figure is not as good as the mass limits set by the other, more direct, experiments reviewed in Chapter 2.

In 1994, the Hubble Space Telescope continued the study of the remains left by the 1987 supernova explosion. With its enhanced power, that telescope has been able to detect three shells appearing as rings not centered on each other. These rings represent the formation of a supernova remnant nebula, although why there are three rings in this pattern is not yet clear. Radio astronomers are searching the nebula for a pulsar, a rapidly spinning neutron star that emits a radio pulse on each rotation, keeping time more precisely than a Swiss watch. Such an object is known to be at the core of the Crab nebula formed in 1006. The remains of the 1987 supernova at present show no sign of a pulsar at its core, but it could be that the neutron star is not spinning rapidly enough. If indeed the "bounce" in the neutrino detection signal was indicative of a neutron star collapsing into a black hole, a pulsar would not be expected: any form of light or radio waves that close to the black hole would not be able to escape. Continued observation may eventually reveal more clues.

Toward a True Neutrino Telescope

There is no certainty that a neutrino telescope will make a fantastic discovery that will once again change our perspective of the universe, but such a telescope may be able to tell us more about the nature of galaxies and other massive objects. Just as scientists studied the Sun's neutrino emission in order to see directly into the core of the solar nuclear furnace, scientists equipped with a neutrino telescope might be able to see directly into the core of a galaxy. Such a telescope could study the neutrino emission of our own galaxy's core and the core of our nearest galactic neighbor, the Andromeda galaxy; a comparison of the two emissions could teach us something about both galaxies. Such a telescope might even help settle the question of whether immense supermassive black holes generate the enormous energy emissions from the cores of active galaxies farther away.

Even though underground instruments have already detected extra-galactic neutrinos, a true neutrino telescope has not yet been built. Such a telescope would have several features that present underground experiments lack. First, it would be better at tracking all the converted particles that result from a neutrino interaction and at reconstructing the original neutrino's direction of travel. Just as a gamma-ray telescope reconstructs the gamma-ray direction of travel by observing the converted electron-positron pair, so too could a neutrino telescope reconstruct the travel direction of neutrinos more accurately if there were some way to precisely measure all the resulting particles. Cherenkov detectors are able to observe only the charged lepton produced in a neutrino interaction, and they cannot produce a good directional reconstruction from this single particle observation. There is, however, a newer underground detector, Soudan 2, that has the ability to track all the charged particles from a neutrino interaction. Second, and much more important, a true neutrino telescope must have a much larger mass. The Soudan 2 detector array, located deep underground in the iron mines of Tower-Soudan, in Minnesota, seems large by most standards: it is made up of 960 tons of steel measuring $17 \times 28 \times 50$ feet. But a true neutrino telescope would have to be 50 to 200 times larger in mass and operate for at least 10 years in order to collect enough neutrinos from our galaxy's core to image it. Third, and finally, the ideal neutrino telescope would be placed high in orbit above the atmosphere or on the surface of the moon where there is no atmosphere. At these altitudes the telescope could avoid high-energy particle interactions taking place in the Earth's atmosphere that also produce many neutrinos. These atmospheric neutrinos come from the interactions of high-energy charged particles that have been bent by the Earth's magnetic fields, and so do not point to any particular place in the sky as their point of origin. On the Earth's surface this problem could be reduced by building a neutrino detector that would be triggered only by very high energy signals, above the atmospheric neutrino's energy, or perhaps only by steeply inclined sideways-traveling neutrinos. Despite these stringent requirements, a neutrino telescope on the Earth's surface is feasible.

Detectors are under consideration that would have dramatically more mass than any built so far. One, called DUMAND, for Deep Underwater Muon And Neutrino Detector, would use the deep ocean as the neutrino detection medium; other, similar experiments would use the water of lakes. Proponents of yet another, the Antarctic Muon And Neutrino Detector Array (AMANDA), propose to detect neutrinos interacting in the Antarctic ice shield. All these experiments would consist of photosensitive arrays placed in water or ice to record the Cherenkov light of leptons from a converted neutrino. They would all be able to detect a coarse direction of travel, just like the water tank experiments that detected supernova neutrinos. AMANDA would not have the spectacular ability to reconstruct a neutrino travel direction that a Soudan 2

A neutrino interaction recorded in the Soudan 2 detector. The detector has recorded the tracks of two charged particles produced by a neutrino interacting with an iron nucleus. With its superior fine-grain tracking abilities, Soudan 2 can reconstruct the tracks of all the particles produced in a neutrino interaction, permitting a precise measurement of the neutrino's initial direction.

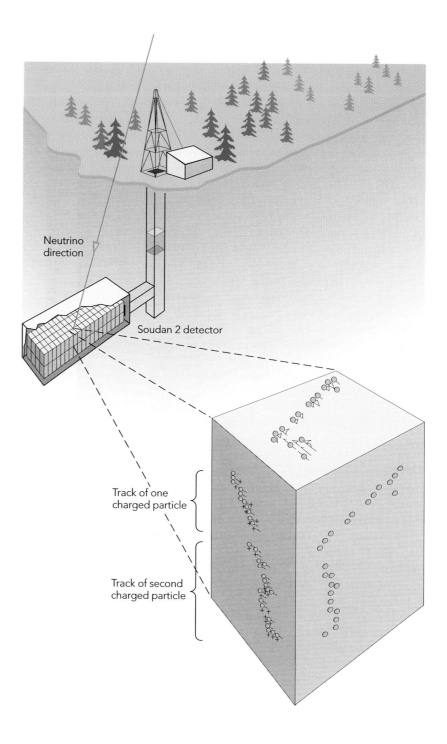

Neutrino direction

Soudan 2 detector

Track of one charged particle

Track of second charged particle

type detector has, nor would it be in high Earth orbit away from atmospheric neutrinos, but it would have, at little cost, the very large mass that guarantees many neutrino sightings.

The AMANDA experiment has just begun operation with a small test array. At shallow depths Antarctic ice is not as transparent to Cherenkov light as the experiment requires, but the project's designers have overcome this difficulty by drilling holes in the ice and placing their photosensitive detectors more than a mile underground. At this depth the ice has been compacted so tightly that it has become transparent. The small array is just beginning to reveal signals from different types of neutrinos. If this array can be expanded to its final planned dimensions of a half-mile on each side, and if physicists can overcome the difficulties of working in the extreme cold of the south pole, it will be a functional neutrino telescope.

Despite present difficulties, we have reason to believe that eventually our vision of the universe will be expanded by systematic neutrino observations. Our new vision will owe its existence to the excitement raised by the neutrino observations from the supernova explosion of 1987, which have already advanced our understanding of stellar collapse and provided conclusive proof that neutron stars exist. The scientific lesson of this chapter is that, while it may be easier to continue studying what one doesn't understand by traditional means, a quick and seemingly miraculous understanding sometimes results when an object is studied in some new way.

. . .

Particle physics detectors deep underground were able to see the first neutrino signal from outside our galaxy, produced by the spectacular supernova explosions of a dying star. These detectors were essentially the first neutrino telescopes in operation; their observations allowed many aspects of supernova explosion theory to be directly tested. The observation of further neutrinos from supernova explosions, our galactic core or the core of our closest neighbor the Andromeda galaxy, or even more mysterious objects such as quasars or active galactic nuclei may await future large neutrino telescopes.

The new neutrino observatories have much in common with gamma-ray observatories that detect electron-positron pairs from interacting high-energy gamma rays. Both are new types of instruments, quite different from Galileo's optical telescope, although they serve a similar purpose. Such instruments look at astronomical objects with new eyes. They have provided a wealth of information for understanding these objects, and we hope to have equal success in the future when we look at the universe through a neutrino telescope.

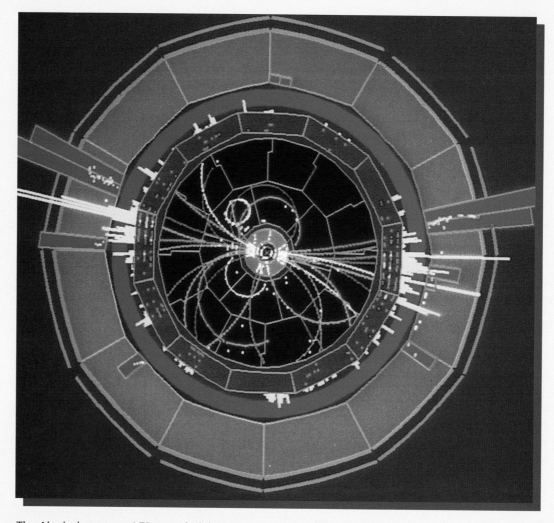

The Aleph detector at LEP recorded this electron-positron collision producing a Z^0 particle. Here the Z^0 decays into two jets of many particles, producing a nice example of a complex event.

CHAPTER 9

Determining the Number of Neutrino Flavors

During a two-year period, from 1975 to 1977, the detection of another generation of leptons and quarks excited physicists at the Stanford Linear Accelerator Center and Fermilab. A third type of electron, heavier than the two types already known, was found at SLAC during an experiment led by M. Perl. The discovery of this new electron, called the tau (τ), mandated a third type of neutrino in order to conserve tau lepton number. That same year, experiments at both SLAC and Fermilab revealed the existence of the bottom quark, part of the third generation of quarks. The top quark was also expected to exist, although to achieve proof of its existence would take much time and effort.

The tau lepton and bottom quark each had a higher mass than the corresponding leptons and quarks of the first two generations. Each was detected through the use of new, high-energy accelerators. With the discovery of these new particles, physicists began to question whether the known quarks and leptons were really all that existed. Perhaps, as each new high-energy accelerator was brought on-line, they would continue to find additional quarks and leptons of even higher mass.

Some particle physicists were convinced that the pattern of quark and lepton generations would continue indefinitely. However, astronomers and cosmologists, through their studies of the remains of the very early universe, would soon be able to tell physicists just how many generations of fundamental quarks and leptons exist. That cosmologists could reveal details about the number and type of neutrinos to particle physicists demonstrates the amazing interconnection between the largest structures in the universe and the smallest, most elusive particles. This, and more, is the story that will unfold in this chapter.

The Beginning of the Universe

The big bang theory of the creation of the universe, mentioned back in Chapter 2, attempts to project the expansion of the universe backward in time. According to the theory, if the projection goes back far enough, then the whole universe will seem to have emanated from one point. What caused this emanation is a mystery that may never be solved. The conditions of the extreme early universe, in the first 10^{-43} second after its creation, are unlike any that we know today from our laboratory tests or astronomical observations; thus any theory attempting to describe these conditions in detail would be mostly speculation. Our best guess is that from the big bang up until the first 10^{-43} second all four known forces of nature were identical and part of a single superforce. After this initial fraction of a second, cosmologists claim to have more intelligent guesses as to what occurred.

At 10^{-43} second after the big bang, gravity had broken off from the other forces. The universe was expanding rapidly, and cooling quickly. At this point matter did not exist as we know it today. Rather, the universe was a great soup of quarks, gluons, leptons, photons, and the W and Z^0 particles that propagate the weak force. Matter was so energetically hot that all the force propagator particles—gluons, photons, and W and Z^0 particles—acted the same. Particles such as the neutrino, which hardly interact in our universe today, were interacting continuously, not only because they had a much higher energy but also because the matter density was so great that particles could not travel very far before encountering other particles.

The interactions of elementary particles, including the neutrino, controlled the development of the very early universe from 10^{-36} to 10^{-6} second. The early universe was very different from today's universe in that these interactions took place at the quark level. This slice of time, from 10^{-36} to 10^{-6} second, brought the universe from the size of a grape to the size of a sphere with the diameter of the solar system. It is at the end of this period that the primordial baryons—protons, neutrons, and other hadronic matter that we know of today—were formed, as quarks and gluons in the cooling universe were finally able to bind to form permanently stable particles. When baryons formed initially, matter and antimatter existed in equal proportion, but because of an unusual property of the previous time period, to be discussed in Chapter 10, matter would soon become dominant.

Our early universe was still small and hot even after the quarks and gluons became bound. At this time only the electro-weak interactions, involving the photon as well as the W and Z^0 particles, remained in equilibrium with matter. What this means is that processes such as

$$p + e^- \leftrightarrow n + \nu_e$$

$$e^+ + e^- \leftrightarrow \nu + \overline{\nu}$$

and even beta decay,

$$n \leftrightarrow p + e^- + \overline{\nu}_e$$

or any other process that involves a neutrino could go in either direction. In the conditions now presiding in the universe, the principle of energy conservation normally allows the reactions to proceed only in the forward direction as indicated by the right arrow. In the very early universe, however, the particles were so energetic that the particles to the right of the arrow could be converted to those on the left. As the universe continued to expand, all the particles cooled, and at a hundredth of a second after the big bang the energy of the average particle fell below 1 MeV. At this point neutrino interactions "froze out" and were no longer in equilibrium. After this cooling, these reactions, involving neutrinos, could proceed only in a single, irreversible direction.

All the neutrinos created after this time are still propagating through the universe. If our understanding is correct, our universe should have close to 1500 neutrinos per cubic inch. Such a high density of neutrinos could easily account for a substantial amount of the missing dark matter in the universe *if* the mass of the neutrino were

The energy of the expanding universe, and thus its temperature, decreased rapidly in the first few minutes after the big bang. Noted on the graph are some interesting points at which certain forces separated or particles froze out. At approximately one second after the big bang, matter and antimatter annihilated, producing more energetic photons and giving a boost to the photon energy temperature ($T\gamma$). At 10^{12} seconds, matter's temperature (T_{matter}) sharply declined when photon interactions decoupled.

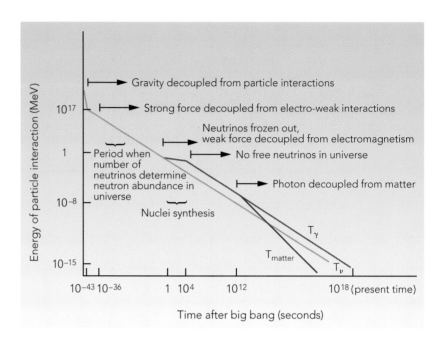

nonzero, as discussed in Chapter 2. This neutrino background, permeating all of space, is analogous to the 2.7-degree-Kelvin microwave background that has its source in photons from the same period. For the neutrino background, however, the temperature is expected to be only 1.9 degrees Kelvin, which means that the average neutrino now has an energy of only 1/100,000 eV. The background temperature of neutrinos is lower than that of photons because neutrino interactions froze out of equilibrium before electron-positron annihilation occurred. The blackbody photon energy spectrum we see today comes almost entirely from electron-positron annihilation. Neutrinos could not get any energy from this annihilation because they were essentially noninteracting with matter at the time it occurred.

Unfortunately, primordial neutrinos have such a low energy that no known technique is capable of measuring them, although actual measurements of the neutrino density or average energy would answer many questions about the early universe and allow current theories to be checked for accuracy. Physicists have been able to detect neutrinos coming from the core of our Sun because these neutrinos have a much higher energy than the primordial neutrinos and hence a larger cross section for interaction. Even easier to detect are the highly energetic neutrinos from a supernova explosion, as we saw in Chapter 8.

After the neutrinos froze out, the next period of growth for the universe was characterized by the synthesis of nuclei. Protons and neutrons became bound together to form hydrogen nuclei; neutrons joined hydrogen nuclei to make heavy hydrogen nuclei; and protons joined with hydrogen nuclei to make helium nuclei:

$$p + n \rightarrow {}^2H + \gamma$$

$${}^2H + n \rightarrow {}^3H + \gamma$$

$${}^3H + p \rightarrow {}^4He + \gamma$$

Nuclei synthesis also occurred through additional interactions, but those are less significant for our discussion. At this time photon interactions had not yet fallen out of equilibrium, so the electrons were still too energetic to be captured by nuclei to make neutral atoms.

At the high energies of the early universe, the interactions of baryons and nuclei were completely different from the processes of nuclei synthesis that take place today in stars. Just after the big bang, these baryons and nuclei were energetic enough to be in electromagnetic equilibrium: the reactions went backward just as readily as they went forward, so the neutrons were replenished just as fast as they were consumed into nuclei. Eventually, about one second after the creation of the universe, the temperature of these particles cooled, the photon interactions froze out of equilibrium, and the neutron interactions did so as well. Now permanent nuclei were being built from the interactions of free protons and neutrons, still the most abundant particles. Because these new particles finally had a sufficiently low energy, once united they were able to stay tightly bound together. It was George Gamow, a theoretical physicist, who in the 1950s first investigated the question of element production in the early universe, and it was Fermi and Tony Turkevich who performed the first experiments to verify the hypothesized reaction rates. Neither of these scientists foresaw that such experiments could have given them new insight into particle physics.

The free neutron is unstable and has a half-life of about 15 minutes. This instability has an important effect on the abundance of helium and hydrogen in the early universe: nuclei can continue to form only as long as neutrons are available. After several neutron half-lives, or about one hour, all free neutrons would have either decayed away or been captured, and nuclei synthesis in the early universe must have come to an end. It should be pointed out that nuclei made up mostly of protons without many neutrons are not stable and quickly decay away. The most stable elements are those that have equal numbers of protons and neutrons; this circumstance has something to do with the strong nuclear force, but the details are not fully understood theoretically.

The abundance of helium and hydrogen depends not only on the half-life of neutrons, however, but also on the initial abundance of these particles. That initial abundance was determined during the period in the expansion of the universe when the neutrino interacted vigorously with other particles, before the neutrino freeze out. The number of neutrino flavors available to participate in particle interactions determined the total number of neutrons available to feed nuclei synthesis, and therefore the relative abundance of the elements, as the following argument reveals.

Before the neutrino froze out of equilibrium, the electrons and positrons were still very energetic, and through the electro-weak interaction involving the Z^0 particle they were capable of producing higher-generation leptons. The more types of neutrinos in participating elementary particle interactions, the more protons could be converted into neutrons through reactions such as these:

$$\overline{\nu_e} + e^- + p \rightarrow n$$

$$\overline{\nu_\mu} + \mu^- + p \rightarrow n$$

$$\overline{\nu_\tau} + \tau^- + p \rightarrow n$$

If the tau-type neutrino did not exist, for example, then there would be one less type of interaction that converts protons to neutrons; while if there were more types of neutrinos, then there would be more ways to produce even more neutrons. The more neutrons present, the more helium-4 would be yielded. As the number of neutrino types becomes very large, all the free protons are converted to neutrons; a very early universe with many neutrino types would yield 100% helium-4 with no hydrogen nuclei. (An additional factor is the time gap between the neutrino freeze out and the point at which free neutrons had cooled sufficiently to bind in nuclei. The existence of the gap gave the free neutron time to decay and reduced its abundance for helium-4 production.) The final ratio of helium-4 to hydrogen is thus very much dependent upon the exact number of neutrino flavors that participated in elementary particle interactions during the neutrino-dominated expansion period of the early universe. If the number of neutrino flavors was greater than three, then there should be mostly helium-4 remaining from the early universe and almost no hydrogen.

Helium-4 was the most abundant *heavy* element in the early universe, because it was the stopping point in the big bang nuclei synthesis chain. All mass-5 isotopes are unstable and quickly decay; the heavy elements we now see in the universe exist because later on stars would be able to concentrate low-mass elements in their centers and get them to interact copiously. The stellar nuclear furnace can produce large amounts of heavy elements by keeping nuclei at high density and high temperature much longer than the rapidly expanding and cooling universe just after the big bang did.

Theoretical models of the early universe showed how the number of neutrino types would affect the primordial production of elements such as helium-4. If scientists could determine the ratio of helium-4 to hydrogen in the early universe, it might be possible to place a limit on the number of neutrino generations. Although many cosmologists independently discussed how the number of neutrino flavors affected the development of the early universe, one especially worthy of mentioning is P. J. E. Peebles of Princeton University. In 1966, he showed that two neutrino types alone could not account for the known amount of helium-4 and suggested that there could be more types of neutrinos. Thus astronomers might be able to measure an important particle physics parameter by studying a diverse number of stellar objects from which they could determine the primordial ratio of helium-4 to hydrogen. Stars in the later stages of their life cycles would not yield this ratio because they have, over the course of time, burned their hydrogen and helium for fuel. Instead, astronomers had to locate objects from far, far back in time.

Galaxies and the gas clouds surrounding them turned out to be the best objects to study for this purpose. The two Small Magellanic Clouds close to our own galaxy are low-mass galaxies that are believed to resemble the first gas clouds from which all galaxies were formed. Each of these galaxies has been found to consist of about 25% helium-4 and about 75% hydrogen, with a very small contamination of other heavy elements. Our own galactic core is 29% helium-4; the Orion nebula, which is an interstellar cloud within our own galaxy, is 34% helium-4. These last two objects are not considered good candidates for measuring the true primordial ratio of helium-4 to hydrogen, because they contain old stars that formed out of the gas cloud, changing its original composition. After taking many measurements, astronomers concluded that the primordial universe had three times as much hydrogen as helium-4.

Real progress in this field came in 1977, when, working independently, Gary Steigman and David Schramm studied again how the concentrations of helium-4 and hydrogen would be affected by the exact number of neutrino types. Steigman and Schramm made the first detailed numerical calculations for the observed concentrations and determined that there could be no more than five neutrino types. Furthermore, measurements favored only three types. Theirs was the conclusive work that finally linked particle physics with cosmology through the most elusive elementary particle of all, the neutrino.

That the study of astronomical objects from our vantage point in the present universe is capable of providing a window into the very distant

Gary Steigman in the center and David Schramm on the right, photographed mountain climbing in the Rockies after a conference at which they discussed the cosmological theory of how the number of neutrino flavors could be determined from astronomical observations.

past is a marvel of modern science. That the study of these same objects can also illuminate the subatomic world of elementary particle physics is still more to be wondered at. Such studies reveal the underlying structure of particles in the universe around us. They prove to us again and again that physicists can help astronomers better understand their many unusual observations and that astronomers, as well as scientists from other fields, can aid physicists in their search for a deeper understanding of the universe.

Measuring the Number of Neutrino Species

After the discovery of a third generation of quarks and leptons, scientists began to doubt whether these three generations indeed represented all the fundamental particles. Many physicists worried that after new accelerators able to achieve higher energies were built, they might find even more generations. Particle physics theories provided only one limitation on the number of generations: that number had to be less than seventeen. Although the standard model of particle physics correctly described all the known particles, and even a few as-yet-unseen ones, it provided no clues to just how many generations of quarks and leptons there might be. Today's standard model does provide the important relationship that the number of lepton generations *must* equal the number of quark generations, as G. t'Hooft first demonstrated in 1979. This conclusion means that the discovery of either the tau lepton or the bottom quark would be sufficient proof that a complete third generation existed.

If they themselves couldn't predict the exact number of quark and lepton generations, physicists wondered, then how could an astronomical observation do so? Even cosmologists questioned their own ability to impose a strict limitation on the number of generations. Particle physicists needed to measure the number of neutrinos directly, in an accelerator experiment. Otherwise, the astronomical predictions would remain an unconfirmed experimental result, and as such they could be quite wrong, as we have seen many times before in this story of the neutrino. It was the scientists working at CERN who took up the challenge.

After the successful observation of the W and Z^0 particles, CERN embarked on a new project called the Large Electron Positron (LEP) collider. This machine was to be built in the largest accelerator tunnel yet dug at CERN, and its beam of particles would be fed by the pre-existing accelerator complex. As the name implies, this new accelerator caused beams of electrons and positrons to collide in a manner similar to the proton-antiproton collider described at the end of Chapter 7. Making electrons and positrons collide was not a new idea, but what was new to this project was that physicists would adjust the energy of the accelerator to be right at the mass of the Z^0. At this energy, the accelerator would produce thousands of these particles per day, allowing a

An aerial view of CERN in Geneva, Switzerland, with the location of the accelerator tunnels drawn in white. Lake Geneva can be seen in the upper left corner.

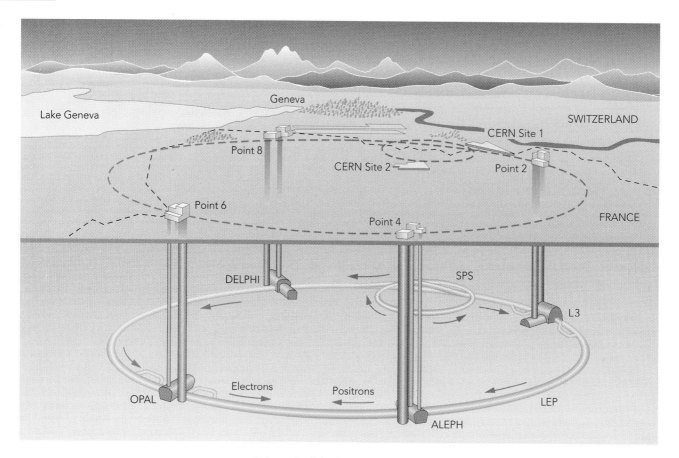

Schematic of the CERN accelerator complex and the LEP collider.

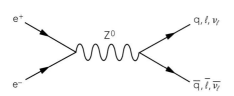

The decay of the Z^0 particle created in an electron-positron collision always produces a pair of antiparticles, but these may be a quark-antiquark pair or lepton-antilepton (ℓ) pair, of which some are neutrino-antineutrino pairs. The members of a pair may be of any flavor or color as long as both are of the same flavor or color.

detailed study of this weak-force propagator particle. The Z^0 particles produced would then decay into any kind of quark-antiquark or lepton-antilepton pair. Some events even produced neutrino-antineutrino pairs that completely evaded detection.

Although the data collected were analyzed in many ways and for many purposes, I will focus on the final, direct measurement of the number of neutrino species. While the beam of electrons in LEP rotates in one direction, the beam of positrons rotates in the opposite direction in the same beam pipe. Although the two beams share the same pipe, they are separated from each other for most of their travel around the accelerator's 16-mile circumference. Magnets steer the beams and bring them together at four points around the ring, where the collisions take place. There is one detector at each of the four collision points, each designed to study the interactions taking place there, and specifically the Z^0 decay. The beauty of this design is that each detector is slightly different from the others, so that physicists obtain a variety of perspectives on the physical phenomena occurring. One detector, called Delphi, provides detailed particle identification of all the mesons and baryons produced, the most difficult task. Another, called L3, has excellent energy resolution for events producing electron-positron pairs from the electron-positron collisions; it also has a good muon-tracking system. The other two, Opal and Aleph, are more conservative, general-purpose detectors. Each of these four detectors provides a computerized electronic readout of the passing particles.

Whereas in the early days of particle physics one or two physicists could perform an experiment on their own, today 300 to 500 physicists work on an experiment as a team. The experiments have grown so large and become so complicated that many of the detector elements are built and operated by separate teams of physicists from one or more universities. These teams are now so big that it is hard to say whether one or a few people are responsible for an experiment's success or failure. One of the official leaders of the Aleph experiment, Jack Steinberger (previously mentioned in Chapter 5 as participating in the neutrino experiments),

The assembly hall of the detector for LEP experiment L3, one of four experiments at CERN that directly measured the number of neutrino flavors. The large number of people in the photo indicates the scale of collaboration needed to perform such an experiment, while the size of the people relative to the equipment indicates the immense size of the experimental apparatus used in modern particle physics. This photograph was taken when construction of the experiment was just beginning.

A view of half the barrel of lead glass crystals that surrounds the interaction regions in the OPAL detector. These crystals detect the outgoing particles by looking at the Cherenkov light produced as particles pass through the crystal.

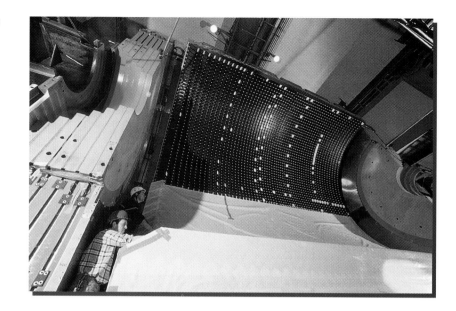

notes that it is all the physicists together who make each experiment work and who are responsible for the accuracy of the final results. Now, instead of depending on the creativity of one or a few individuals, the scientific community must encourage and even cherish the talents of people from all over the world who come together to build these large experiments. While it is true that physicists from different institutes sometimes cannot communicate well in a single language, this predicament provides a splendid opportunity for the physicists of the world to show that they can work together instead of competing as in decades past, when one team in the United States would try to "beat" another in the Soviet Union, China, or Europe. Cooperation becomes especially important when large experiments, and therefore large budgets, are needed in order to advance scientific knowledge.

The LEP experiments directly observe the decay of the Z^0 into charged particles; from the energy of the decay products they obtain a value for the Z^0 mass. What they are actually observing is the spread in energy or mass of the Z^0's decay, called a "width." Even though we say that the mass of the Z^0 is 91.2 GeV, this value actually refers to the center of a broader distribution. Some Z^0's produced are slightly heavier or lighter than this average mass value. The width of the distribution, called the decay width, is significant. This observed decay width does not include any measurements from the decay of the Z^0 into neutrino-antineutrino pairs because such pairs completely avoid detection. However, there is another way to measure the mass distribution of the Z^0: instead of measuring the energy of the particle's decay, we measure the energy required to produce the particle, or its "production width." If a

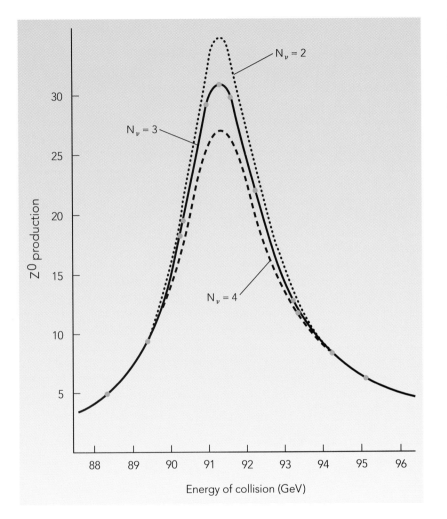

Data from the measurement of the Z^0 yield for slightly different energies of electron-positron collisions. The three curves show the values expected theoretically for different numbers of neutrino flavors. The data clearly show that only three types of neutrino couple with the weak force.

measurement of the Z^0 width for production could be observed, then it would be independent of the neutrino because the accelerator itself has only electron-positron collisions occurring. Physicists adjust the energy of the accelerator around the mass of the Z^0 and measure the Z^0 yield for slightly different energies of electron-positron collisions. As the energy of collisions gets closer to the mass of the Z^0, the number of Z^0 particles increases. Physicists can thus obtain a true measurement of the width of the Z^0 particle produced in this manner. The difference between the observed decay width and the total production width, measured by varying the energy of the accelerator, gives the exact number of neutrino-antineutrino species that couple to the Z^0. So, in essence, by measuring the width of this particle's production we can learn how many species of neutrinos are coupling through the weak force. Numerical studies from this data tell us that there are exactly three, and only three, types of neutrinos.

This finding is in splendid agreement with the astronomical observation of the hydrogen-to-helium ratio of the universe. What the astronomers had told us about the number of neutrino species, physics experiments here on Earth confirmed. This result justified a new confidence in cosmological models of the universe and the power of physics. Three hundred years ago Galileo turned his telescope to the stars and confirmed that Newton's laws of motion on Earth also held true for other celestial bodies; similarly, modern astronomers studying remote stars know that the quantum mechanical laws of atomic spectra observed on Earth also apply on the other side of the universe. The implications are astounding. The lessons we learn today in the laboratory can be applied to the physics occurring at the very beginning of the universe—millions and millions of years ago.

Admittedly, one can imagine hypothetical situations that would permit additional generations of neutrinos, but these theories require either that the additional neutrinos be highly unstable—very short lived—or that they be very massive. In either of these cases the additional neutrinos would have made only a negligible contribution to the particle interactions occurring in the early universe. But we know that the three generations of neutrinos that exist are *not* unstable, and all seem to be very lightweight. If other forms of neutrinos existed, they would be distinctly different from the neutrinos of the first three generations—a seemingly unlikely scenario.

The Quest for the Last Quark

By 1994, physicists had great confidence in the theory of particle interactions, commonly called the standard model. Only one problem remained: Where was the top quark? The extreme skeptic argued that if the top quark could not be found, perhaps the standard model was not as good a theory as everyone thought. To find the top quark, measure its mass, and study its decays would put the finishing touches on the standard model. This time in the history of particle physics, everything was to occur as predicted.

The highest-energy accelerator in the world today is the Tevatron at Fermilab. It is a proton-antiproton collider very similar to the CERN SPS collider, but it accelerates beams to a higher energy of 900 GeV. (An energy of 1000 GeV is called 1 TeV; the T stands for "teva," from which the accelerator gets its name.) There are two collider experiments, each located at one of the four regions where the two beams intersect; one is called the Collider Detector Facility (CDF) and the other D-Zero. These detectors were built in the very early 1980s, while the accelerator was being upgraded to double the energy of the original machine and, following in the footsteps of CERN, an antiproton source

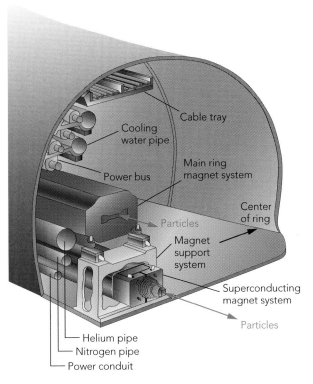

Cable tray

Cooling
water pipe

Power bus

Main ring
magnet system

Center
of ring

Particles

Magnet
support
system

Superconducting
magnet system

Particles

Helium pipe
Nitrogen pipe
Power conduit

Inside the Fermilab accelerator tunnel. The blue and red magnets on the top are the older main ring initially built in the 1970s, while the bottom accelerator ring is the newer superconducting Tevatron accelerator. The Tevatron takes protons and antiprotons to an energy of 1000 GeV; at this energy they collide, at the experimental hall, and produce new, very massive particles, such as top-antitop quark pairs. The superconducting magnets need to be cooled to almost absolute zero, by liquid nitrogen and helium, to achieve their superconducting property of no resistance so that the accelerator is inexpensive to operate. Cable trays bring electrical power to the magnets, and water is also piped in to keep the magnets cool.

was being built to transform the machine into a collider. Finding the top quark at these new high energies was not going to be easy, since these higher-energy collisions would produce many more secondary particles than any experiment before. Thus the task would demand significantly improved experimental apparatus for particle tracking and energy measurement. To achieve the necessary level of detail would require a lot of time and effort on the part of many experimenters.

The experimenters needed a clean way to trigger the apparatus when an event signaling a possible top quark occurred. A top quark would always be created together with an antitop quark, but the detectors could observe neither quark directly—instead, they would search for the products of their decay. By 1990 physicists were certain of one thing—that the top quark was heavier than the W particle, so it should preferentially decay by coupling to a W particle and the other quark in the third generation, the bottom quark. Since both the top and antitop quark would couple this way, there would be two W particles, each

A Feynman diagram of the top-antitop quark decay mode that was used to select events as top-quark candidates.

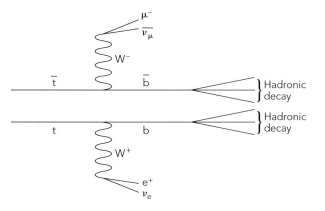

decaying into a weak-force doublet of two quarks or two leptons from the same generation. Electrons and muons were the easiest particles for either of these two detectors to measure confidently, since the detectors could identify these particles unambiguously and measure their energy with a high degree of accuracy. Therefore, the easiest events to detect were those in which one W coupled to an electron and its corresponding neutrino and the other coupled to a muon and its corresponding neutrino. Thus a "clean trigger" would find those events producing a single high-energy electron and a single high-energy muon.

This trigger, although the best for spotting candidate top quark events, was not the most efficient. The W particle has many more ways to couple to quark doublets, which are also produced in the decay of the top quark. To be precise, a W decay is nine times more likely to produce a quark doublet than a lepton doublet. Even though the single muon and electron trigger was cleaner, the "cleanness" would come at the expense of fewer chances of detecting the top quarks. Yet, because this was the most accurate means of detecting the top quark if that quark existed, it was the chosen method for this experiment.

The CDF experimenters announced in 1994 that they had clear evidence of events producing one muon and one electron, six events in all. They gave the mass of the top quark as 174 GeV/c^2. At that time the other experiment had found no evidence of the sought-after particle, so the collection of data and its analysis continued. By the spring of 1995 CDF had seen 25 pristine events indicative of the top quark, and D-Zero had also, finally, seen similar events. The CDF experiment currently places the mass of the top quark at 176 GeV/c^2; the D-Zero experiment estimates a mass of 200 GeV/c^2 from substantially fewer reconstructed top quark events. With these newest measurements, the top quark's existence is now firmly established, although refinements to the mass measurement are still needed, as is a study of the different decay modes.

When the experimenters searching for the top quark decided to use the special trigger of one muon and one electron, they were not only sacrificing the number of events they could observe, but also compromising their ability to measure the particle's mass. When the W particle couples to a lepton pair, such as a muon and its neutrino, some unobserved amount of energy escapes with the neutrino. When the W couples to a pair of quarks, none of the energy escapes unobserved, but the decays have a much more complicated structure. These events produce groups of particles that stay close together, commonly referred to as jets. Although both detectors are capable of identifying and measuring jets, they have to be able to distinguish top quark events from many

This image taken at the Tevatron accelerator captures the particle tracks of a proton-antiproton collision that produced a top-antitop quark. The two beams of protons and antiprotons are at the very center perpendicular to the photo; the white dots are observed hits in the central tracking detector, while the colored lines are the best tracks that computer algorithms found using the detector's response. Outside this ring is a layer of energy-measuring devices, drawn as pink or blue boxes; their height is proportional to the amount of energy deposited there. Muons can even penetrate the energy-measuring devices, and their tracks are shown as different colored "+" marks. The red arrow indicates where missing unobserved energy, carried by neutrinos, has gone.

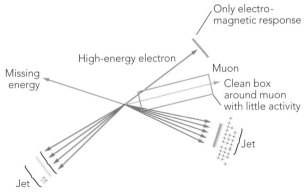

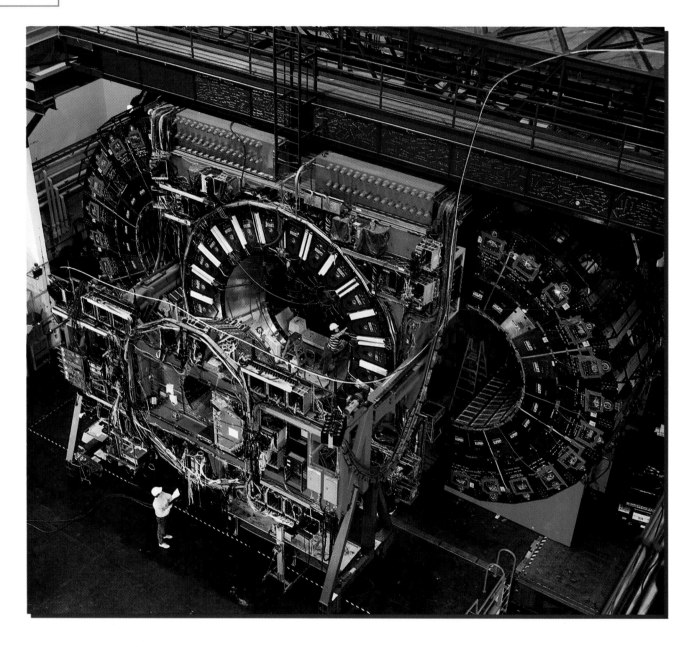

The CDF detector, the detector that observed the first top-antitop quark events. Here the detector is seen partially disassembled, with its two arcs, which are the energy-measuring devices, separated from each other. When pushed together, they surround the central tracking chamber.

other types of interactions that have similar jet structures but do not produce top quarks. The CDF experimenters became capable of doing just that. They were able to identify events producing two bottom quarks, each of which decayed into two jets, indicating a candidate top quark event. There are no neutrinos produced in the initial stages of these four-jet events, and so physicists may make a much better determination of the mass of the top quark. When the CDF detector searched for these events, it found many more top quark candidates—about a hundred were identified in the experiment's 1995 announcement. The experiment thereby achieved a better statistical sample of top quarks and a better study of the mass.

In retrospect, both types of analysis were necessary. The "clean trigger" on events with a single muon and a single electron convinced physicists of the existence of the top quark, while the four-jet events provided the most accurate measurement of mass. Therefore, the CDF-estimated mass of 176 GeV/c^2 is the better of the two measurements.

In all the searching for different particles, from the top quark to the other quarks that came before, one unknown has always been the mass of each new particle. No theory has been able to predict the mass of quarks or leptons. Why do certain particles have different masses? This question, to be further considered in the final chapter, still haunts particle physicists today.

. . .

We have seen in previous chapters that chemistry, geology, statistics, and even biology have contributed significantly to experiments in particle physics. This chapter has stressed the connection between elementary particle physics and astronomy. Unlike astronomy, which is mostly an observational science, particle physics is an experimental science that permits a direct study of matter and its formation. While the study of astronomy can help us to understand particle physics, physics is in turn a strong and necessary foundation for any real understanding in astronomy. All scientific fields benefit from interdisciplinary study, and many of our new technologies, advances in medicine, and philosophical discussions about the world in which we live depend upon such studies, which help satisfy a desire within us to know the unknown and to discover just how this intricate universe came into being.

Neutrino interactions in the very early universe controlled the formation of nuclei. These nuclei formed atoms, then atoms formed gas clouds, and the gas clouds eventually condensed to form stars, which clumped into galaxies. In this way, a small, almost impossible to detect particle had a profound impact on the universe we see. Studies of this most elusive particle interacting in today's modern accelerators have given scientists a better understanding of the particle interactions that must have taken place in the very early universe. The agreement of physics and astronomy in determining the number of neutrino flavors has persuaded scientists that they can trust this new scientific understanding of the universe in the distant past, gained from accelerator experiments. Now scientists are able to make particle physics measurements that confirm the conditions from which our universe developed.

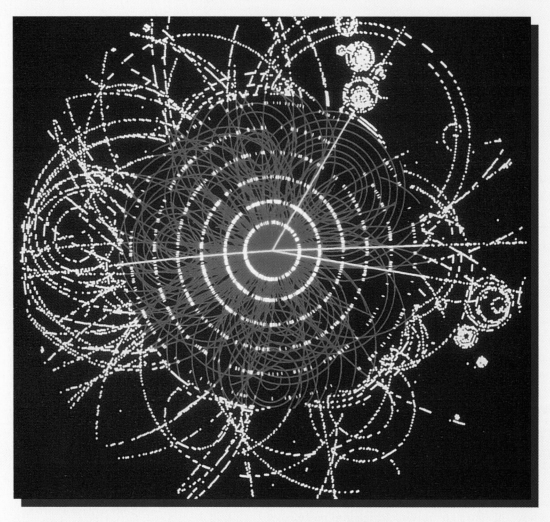

A simulation of a proton-proton collision at 7.5 TeV in a future collider at CERN, producing a Higgs particle. Here this yet-to-be-observed particle is identified as it decays into four energetic muons drawn in yellow; the tracks drawn in red are lower-energy particles.

CHAPTER 10

Investigating the Unanswered Questions of the Microcosm

Although our knowledge of matter and the forces that it is subject to has advanced greatly in the last sixty years, there is still much we do not understand. The variety displayed in quark masses remains a mystery, one that may be solved by a recent theory if it can be proven correct. The strong nuclear force that binds protons and neutrons together is another mystery of sorts, for we do not fully understand its behavior. An additional puzzle is why the matter of our universe is not accompanied by an

A table of elementary particles and forces. The reason the particles are of different masses remains a mystery.

ELEMENTARY PARTICLES

Quarks	Electric charge	Mass (GeV/c^2)	Leptons	Electric charge	Mass (GeV/c^2)
u	+2/3	0.002 – 0.008	e	1	0.511 MeV/c^2
d	–1/3	0.005 – 0.006	ν_e	0	<9 eV/c^2
c	+2/3	1.3 – 1.7	μ	1	105 MeV/c^2
s	–1/3	0.1 – 0.3	ν_μ	0	<270 KeV/c^2
t	+2/3	175 – 200	τ	1	1.780 MeV/c^2
b	–1/3	4.7 – 5.3	ν_τ	0	<350 MeV/c^2

FORCES

Electromagnetism:	γ	Photon	0 mass
Weak force:	W$^\pm$	Bosons	86.9 GeV/c^2
	Z^0	Bosons	90.1 GeV/c^2
Strong nuclear force:	g	Gluon	0 mass

equal amount of antimatter, but CP violation may offer an explanation if a few problems can be solved.

These mysteries of the microcosm of particle physics, once solved, may have implications that reach far beyond physics into other fields, especially cosmology. Scientists are now trying to simulate conditions in the universe just after the big bang, cooking up "soups" of quarks and gluons through the collision of heavy ions. Through such relativistic nuclear experiments, solutions to some of the riddles of particle physics and our universe's early development may emerge.

In the previous chapters the neutrino has served as a guide, opening a window onto the large expanses of particle physics. In this chapter we open the window wider still to look at other important mysteries, regardless of whether the neutrino has a role. The neutrino will still have a special part to play in regard to the question of the origin of the different masses for elementary particles. In addition, it will be interesting to see how the neutrino can interact with the particles that propagate the new superforce expected from the unification of the strong force with the electro-weak force.

The Origin of Different Particle Masses

We have seen that matter is composed of six quarks and six leptons. Normal matter, the matter of our day-to-day experience, is composed of the first generation of quarks and leptons. The remaining quarks may be present in other places in our universe but were especially prevalent in the distant past, just after the big bang occurred. These different quarks and leptons have distinctly different masses, but where does this mass come from and what properties of nature cause each particle to have a different mass?

Three forces of nature—the electromagnetic, weak nuclear, and strong nuclear—are propagated by special particles and consequently are certainly *quantum forces*. Two of these, the electromagnetic and strong nuclear, are propagated by a massless particle, while the weak nuclear force is propagated by particles that have mass, the W and the Z^0. The fourth force of nature, gravity, has no known quantum formulation, but it does have a proposed force propagator particle, given the name graviton. Currently we are uncertain of its existence since it eludes direct observation. If scientists were to observe this graviton, then we would know that gravity is truly a force with a quantum formulation. If the graviton exists, then it too must be a massless propagator since if the graviton had mass, gravity could not interact over great distances. Why does nature find it necessary to produce distinctions between the forces and between the masses of their propagator particles?

Although physicists do not completely understand how particles acquire mass, they are exploring two good theories.

Peter W. Higgs, a Scottish theorist, proposed the first of these theories in 1964. His theory of how particles acquire mass is now considered an essential part of the standard model of particle physics. Higgs suggested that there is a field, today called the Higgs field, that permeates all of space, comparable to the electromagnetic fields around power lines at high voltage. Electromagnetic fields are the lines along which two charged particles propagate their force; in contrast, the Higgs field couples with particles, tugging on quarks and leptons in such a way that some types end up with higher mass than others. Thus the mass of a particle depends on how well it couples to the Higgs field.

An analogy from politics might help to illustrate the effect of the Higgs field. Imagine a large reception at the Capitol building in Washington. All members of Congress and all major political activists are in attendance. Members of each of these groups are uniformly scattered around the room. Further imagine that Congress has a Democratic majority. If Bill Clinton enters the room and walks toward the speaker's platform, many Democrats will stop him to shake his hand; all the handshaking slows him down. In effect, the many handshakes will impart to Clinton an apparently large mass. There are appreciably fewer Republicans, so if George Bush enters the room, he will not be slowed down by as many handshakes, and hence Bush will acquire less mass than Clinton. But if the still-popular Ronald Reagan enters the room, he will be slowed down considerably more than his party's more recent president, imparting to Reagan a very high apparent mass. Just as

Peter W. Higgs, who in 1964 proposed a revolutionary explanation for the origin of the different masses observed in quarks and leptons. Although his theories have yet to be proven experimentally, Higgs's theories are the most talked-about physics of our times.

different presidents from the same party seem to differ in popularity, not all particles of the same type necessarily have the same coupling strength with the Higgs field. Now, if Ross Perot, who is very unpopular with members of Congress, enters the room, very few will wish to shake his hand and he will have no problem dashing through the crowd. Perot's experience is similar to a neutrino's behavior in the Higgs field. The neutrino has a characteristically small coupling with the field, which leaves it with an extremely small mass or no mass at all.

The model does not explain the physical origins of this proposed, all-permeating Higgs field. Since the Higgs phenomenon is a field, it should have a particle that propagates the field. Such a particle would be expected to have an extremely high mass. Finding the Higgs particle, a discovery that would help to confirm that mass is generated by the Higgs field, was one of the compelling reasons for building the Superconducting Super Collider (SSC), plans for which were canceled in 1994. Currently, CERN is planning a new collider, called the Large Hadron Collider (LHC), that will take advantage of existing accelerators and be located inside the already existing LEP tunnel. It will have just a third of the energy that the SSC would have had, but is expected to be just as prosperous for science and may well be able to find the Higgs particle. Because the LHC is the result of international cooperation, its funding is not dependent upon any one country and its chance of completion is good.

Theoretically, the Higgs field is still present even when no particles are present. Thus the Higgs field could be giving rise to particle-antiparticle pairs generated in the vacuum of space. Because the ground state of the Higgs field is not zero (unlike the electromagnetic field of QED), particles created out of the vacuum can propagate and interact. So, when physicists talk about understanding the vacuum of space, they are also talking about understanding how particles of differing mass and their antiparticles are created. The Higgs field could offer an understanding of the vacuum of space.

The other theoretical explanation of mass is found in what is called the *compositeness theory*, which states that quarks and leptons are composed of sub-subparticles called preons. According to this theory, quarks and leptons are just different excited states of the ground state coupling of the preons.

Imagine that the up and down quarks are composed of preons, while the electron and its accompanying neutrino are also composed of some smaller preon particles. According to the compositeness theory, when the preons are excited by a high-energy particle interaction, they would reconfigure into the heavier strange and charm quarks or into muons and their accompanying neutrinos. Nuclear and particle physics provide ample evidence for a similar effect: identical combinations of

The LEP accelerator tunnel, where the future Large Hadron Collider (LHC) will be built to search for the Higgs particle.

quarks form different particles with different mass states. For example, the proton and delta (Δ^+) particle are both composed of two up quarks and a down quark, but because the delta's quarks are bound together in excited orbits, they produce a particle of heavier mass, though much shorter lived than the proton. It may help to refer to the figure on page 96, where the organization of quarks into observable particles was first presented; note the different kinds of particles composed of the same quark combinations.

Rutherford probed the structure of the atom by scattering alpha particles off a foil of gold; this investigation led to his discovery that the positive charge in an atom is concentrated in a nucleus surrounded by orbiting electrons. Later, physicists at the Stanford Linear Accelerator probed the structure of the atom even more deeply by scattering electrons off protons; by this means they identified the subatomic particles called quarks. Today, a special accelerator located in Hamburg, Germany, is performing a search for an even deeper substructure. Called HERA, this accelerator causes counterrotating beams of electrons accelerated to 28 GeV to collide with protons of 820 GeV. HERA is exploring the structure of the quark, hoping to determine whether there is any evidence for preons. Begun in the early 1990s, this experiment probes the proton in the smallest dimensions currently possible; it will continue to search for smaller substructures for many years to come.

If a quark has only one charged preon, however, then probing it with electrons will not reveal a substructure even if one is present. Another way to investigate the composition of quarks is to scatter high-energy neutrinos off protons. The neutrino is a more perfect, pointlike particle and can probe more finely than the electron. It would not be as sensitive to charged particles as the electron, but because it interacts only through the weak force it is sensitive to all weakly interacting subcomponents, even those without charge.

Understanding the Strong Nuclear Force

The strong nuclear force is the force that binds protons together in a nucleus, and it is the same force that binds quarks together to form protons, neutrons, and other hadrons. The complexity of the force, and the difficulties it presents to those who wish to thoroughly understand it, are suggested by a review of its propagator particles. The long-range component does most of the binding work in a nucleus, while the short-range component holds the quarks together in a proton or neutron. Gluons are the propagators of the strong nuclear force; they act like a nuclear glue to bind the quarks together. While each quark has a single color state, a gluon has dual color states that permit it to bind together quarks of different color. Two baryons bind together by exchanging a virtual meson, the neutral pion (π^0) for instance. When two protons exchange a virtual meson, the virtual meson gets one quark from each proton. As the meson is exchanged, each quark simultaneously travels in the opposite direction, and each quark "sees" an antiquark going by when the two quarks pass each other. This virtual meson exchange has the same property of propagating the strong nuclear force as the gluon exchange because, like gluons, mesons carry two colors, in this case a color and its exact opposite color (while gluons can carry any two colors). However, between quarks in a proton, virtual meson exchange plays less of a role.

There are some important differences between the photons of electromagnetic interactions and the gluons of strong interactions. Photons can be free, while gluons must always be connected to a quark. Photons cannot interact with other photons, while gluons can interact with other gluons as well as with quarks. The strong nuclear force allows one more degree of possible interaction, making for a more complicated phenomenon for which it is more difficult to develop a proper theory.

Our understanding of the strong nuclear force took one of its most recent advances in 1979, when four experiments at the electron-positron collider at DESY Germany observed clumps of particles staying together in jets. Some physicists may justifiably argue that scientists working on two earlier experiments at CERN found the first signs of jet production in 1973 at the Intersection Storage Ring (ISR). However, the CERN experiments did not actually see jets until 1979, when an improved Axial Field Spectrometer experiment was able to achieve their detection. All these experiments observed the existence of jets, whether by high-energy interactions of 13 GeV electron-positron collisions as at DESY, or by the head-on collision of two protons of 26 GeV each as at the ISR. The UA2 experiment dramatically confirmed this discovery several years later; the spectacular events it created produced back-to-back jets, many with over 20 particles in each jet.

Jet formation provides excellent insight into the nature of the strong nuclear force. In the theory of quantum electrodynamics (QED) each charged particle is surrounded by its own virtual photon cloud, and it is this cloud that propagates the electromagnetic force. This force falls off with distance (D) in the proportion $1/D^2$. But in the theory of quantum chromodynamics (QCD), the analogous quantum theory for the strong nuclear force, each quark or gluon is surrounded by a gluon virtual field of color. As one quark is pulled farther away from the other, the strong force exerts a restoring force that *increases* linearly with distance. At zero distance apart, the quarks act as if they are free from each other, a behavior referred to as *asymptotic* freedom, but any slight motion that separates the quarks creates a force proportional to the distance of separation, pulling the quarks back together. When quarks are bound inside a baryon or meson, they are almost completely shielded by their partners from any attraction to the quarks of other baryons.

All this explains why free quarks are not observable. If we could take just one quark out of a particle and move it to infinity, there would be an infinite force that would pull the quark back to the particle. Since an infinite amount of energy is impossible, completely freeing a quark is impossible as well.

Before a quark can be pulled very far away from other quarks, the force lines between them break in half, producing a quark-antiquark pair from the vacuum. In a collision of particles such as takes place in an accelerator, a large amount of energy is imparted to a colliding quark. As this quark tries to separate from another quark, the force

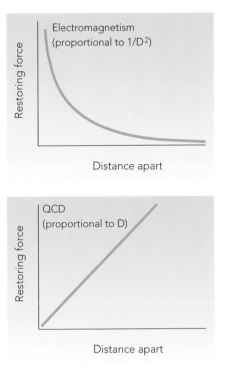

The electromagnetic force of attraction (top) is inversely proportional to D^2 (where D is distance), whereas the strong nuclear force of attraction is directly proportional to D (bottom). One consequence is that a quark completely separated from its partner would experience an infinite restoring force.

lines connecting it to the other quark break apart, and a quark-antiquark pair forms. One of the quarks in the newly formed pair stays with one end of the initial quark and the other stays within the interacting particle. To illustrate this process, imagine two children playing with a long slinky, with one child pulling on each end. The children represent the quarks, and the slinky itself is the strong nuclear force keeping them together. If two more children ask to play, the original two could cut the slinky in half and give the cut ends to the new children. The end result would be that each new child is playing with an end of the slinky attached to one of the initial children. This is exactly how the quark-antiquark pair created from the vacuum attaches to the existing quarks of the collision.

When a lot of energy is applied to such an interaction, this process can occur several times, creating many particles and forming a jet. Yet the process alone does not explain why the particles clump together. The jet is explained by the strong nuclear force, and demonstrates its truly immense strength. Even though the force lines between separating quarks snap and many quark-antiquark pairs form, the strong nuclear force still causes the separating quarks to have a high affinity for one another.

Much is still unknown about fundamental quark interactions mediated by the strong nuclear force. We don't yet know how the force lines snap, and why different types of quark-antiquark pairs are produced. We don't know the relative strengths of the long- and short-range components of the strong nuclear force and exactly which kind of virtual particle exchange contributes, and by how much, to these interactions. These questions present particle physicists with exciting frontiers to explore.

Another frontier is the exploration of the strong nuclear force by heavy ion collisions. Experiments with collisions between heavy ions began at Berkeley in the 1950s and continue today at Brookhaven and CERN. In the near future Brookhaven will have a heavy ion collider that provides head-on interactions between beams of gold nuclei. These experiments smash large nuclei together in an effort to increase the density of nuclear matter. Finding out how high that density can go is just one of its goals, and determining whether the unusual matter that results would have any unique properties is another.

The experiments are designed to create nuclear matter of sufficient density that the quarks and gluons previously bound in protons and neutrons merge into one very massive collection, a quark-gluon plasma in which no baryonic structure remains. Through these experiments, physicists should be able to learn at what density protons and neutrons lose their identity. The quarks and gluons in the plasma will be squished together so compactly that they will act as free particles. Particle physicists expect to learn a lot about the strong nuclear force just from the

creation of this unusual state of matter. Although the loss of proton identity would, in principle, violate baryon number conservation, the identity loss is not an immediate concern because this empirical law of physics is being violated at extreme conditions. Baryon number conservation is not as fundamental as the conservation of energy and momentum, the violation of which would completely halt such theoretical speculation. As Pauli correctly pointed out while wrestling with the conception of the neutrino energy, conservation laws are still a nonremovable cornerstone of all physics. However, baryon conservation is expected to be violated in collisions that take place at an energy that is high enough.

A single heavy ion interaction produces hundreds of charged particles. The image shows one such event, from CERN experiment NA-35, recorded in a special type of optical spark chamber.

Heavy ion collisions achieve high temperatures and densities that approach those in the very early universe. In these conditions a quark-gluon plasma is expected to exist. By studying heavy ion collisions, physicists hope to learn at what energy or density the quark-gluon plasma forms, in order to gain more insight into the early universe, the interior of neutron stars that may have exotic forms of quark matter, and, most important, the interactions of quarks and gluons at high energies.

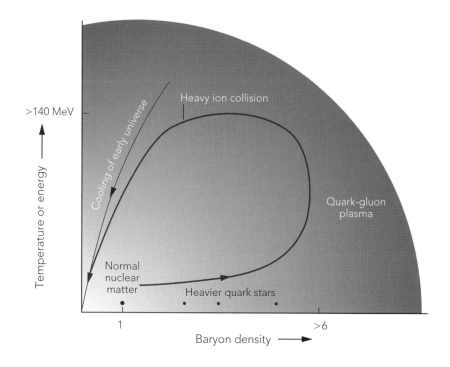

According to the big bang theory, the universe was initially a highly dense quark-gluon soup of nuclear matter, similar to the plasma that the Brookhaven and CERN experiments are trying to create. The "normal" matter of our everyday experience condensed out of this soup. The heavy ion collision experiments provide the perfect opportunity for cosmologists and astrophysicists to study, for brief moments in the laboratory, conditions similar to those in the early universe, something they could never do strictly through observation of the night sky.

Nuclear physicists and astrophysicists are also interested in the heavy ion collision experiments for what these could tell us about the formation of superdense matter, similar to the matter that bizarre stellar objects, such as neutron stars or possibly quark stars, are composed of. If these experiments greatly increase the density of normal nuclear matter, they may show us whether such material at higher densities compacts to form heavy quarks from the second or third generation. They could thus tell us if, at the center of large neutron stars, there may be cores made of heavier quarks, forming strange matter, or if stable stars made only of even heavier bottom and top quarks could possibly exist.

Another mystery surrounding the strong nuclear force is the EMC effect, named after the European Muon Collaboration experiment that discovered this perplexing phenomenon. Like Rutherford's experiment, the EMC experiment used an intense beam of charged particles to probe atomic structure—in this case, a beam of muons to probe quarks. Both experiments aimed their particle beams at simple nuclei, such as hydrogen, as well as complicated nuclei, such as carbon and iron. Like the particles of other scattering experiments, the muon was able to probe the charge distribution, but because the muon, at two hundred times the mass of the electron, is a smaller, pointlike particle, it does not produce the messy interactions that electrons do.

Theoretically, quarks should be completely confined to their proton or neutron particles, regardless of whether the proton is part of a large nucleus or a small one. However, the EMC experiment found that the muons scattered more strongly off the quarks bound inside a proton in a large nucleus than off simpler nuclei such as hydrogen, which contains only a single proton. This evidence suggests that quarks are not confined to protons without hope of escape as was previously thought. It also tells us that large nuclei are somehow sharing their quarks among their protons and neutrons. The experiment has not yet determined where this quark sharing in the nucleus occurs, whether among the outer protons and neutrons or in the center of the nucleus. What we do know is that a better understanding of this curious phenomenon will provide insight into the structure of the nucleus.

Toward a Grand Unified Theory

Physicists generally believe that all elementary particle interactions are in some way related. Based on this belief, and on their success in incorporating the weak nuclear force into the same theoretical formulation as electromagnetic interactions, most physicists also believe that a foundation for a grand unified theory is in place. Such a theory would incorporate the strong nuclear force with electro-weak interactions into one supertheory. Elementary particle interactions do not appear to us at our current low energies to be mediated by a single superforce, but such a force is postulated to mediate interactions at much higher energies, and is a key to understanding the first instant of the big bang.

We have already seen one instance of a unification of forces, in Chapter 7, where I discussed the incorporation of weak and electromagnetic interactions into the first unified theory of particle interactions. The story told in that chapter provides some insight into how a search could be performed for evidence of further unification of the forces.

At low energies—those that dominate the world we live in—weak and electromagnetic interactions are distinct. However, at Fermilab, at CERN's SPS or LEP colliders, and at HERA, the energy scale of interactions surpasses the mass of the Z^0 particle; in the conditions created within these accelerators, the weak and electromagnetic interactions are equivalent in strength. Physicists have inferred that, similarly, at some even higher energy scale the strong nuclear force should approach the strength of electro-weak interactions, and the two types of interactions should be unified. Since the high energies needed to produce such a unification outstrip the energy production of present-day accelerators, how else might a grand unification among forces be seen?

A careful contemplation of the early days of weak-interaction discoveries provides us with two possibilities. First, the weak force and the weak-force propagator, W, had to be invented to explain the decay of the neutron in a nucleus into a proton, electron, and neutrino. Second, an additional, neutral propagator of the weak force, Z^0, was later required to explain observations of neutral-current neutrino interactions. Analogously, interactions between quarks and leptons that cannot be explained by either electromagnetic effects or the W or Z^0 particle exchange might provide sufficient evidence for a particle mediating a grand unified force. Such interactions require only virtual propagator particles, so the discovery of a new particle decay or a new interaction between quarks and leptons might reveal the grand unification without the need to actually create its propagating particle at the extreme energies necessary.

The discovery of any process that couples the proton with the electron, and thus violates the conservation of baryon and lepton number, would show that the strong nuclear force could couple with electroweak interactions. Furthermore, if such interactions occur, then even the neutrino, which is a lepton as is the electron, would interact with the proton and other hadrons. It has been postulated that the easiest way to find such an interaction would be to observe a proton decay that violates baryon number conservation. Obviously, if proton decay does exist, protons must have a very long lifetime, since we know that large amounts of matter—including the Earth and our solar system—have existed for extremely long periods of time. Nevertheless, theorists were optimistic enough to propose the existence of two new propagator particles, X and Y, that would couple quarks and leptons.

Just after these ideas were circulated, the lifetime of the proton based on the best estimates for the age of the universe was placed at 10^{24} years. Since then, several large experiments have been set up to search explicitly for proton decays that violate baryon number conservation. The large underground water detectors that saw the first supernova neutrinos and the Soudan-2 type detectors with their superior tracking abilities, all previously discussed in Chapter 8, were originally built to search for proton decay. No proton decays have thus far been

A Feynman diagram of a quark-lepton coupling through the proposed propagator particles of a grand unified theory. This diagram demonstrates how proton decay might proceed through a virtual particle decay mediated by as-yet-unseen force propagator particles.

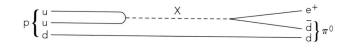

observed that would indicate the existence of the sought-after propagator particles. This lack of results has forced physicists to adjust their estimates of the proton lifetime upward: today the best estimate is that the half-life of the proton is greater than 10^{32} years.

The HERA accelerator—a collider of electrons and protons—is also looking for interactions between quarks and leptons that would demonstrate the existence of the proposed X and Y particles. Current data show clearly that at an energy above the mass of the Z^0, the electromagnetic force is equal in strength to the weak force, just as predicted. It is expected that the strong nuclear force will have a particle similar to the Z^0 such that at an energy above the mass of this particle, the strong nuclear force would join with the electro-weak force. However, before reaching that energy scale our accelerators should produce virtual particle interactions that violate baryon and lepton number conservation, and these would demonstrate how quarks and electrons interact. It is hoped that experiments will eventually discover the first clue to grand unification if, in fact, the forces are really explainable by one grand theory. The real excitement, however, would be in the discovery of the true conservation laws under grand unification.

Investigating the Mystery of CP Violation

We saw back in Chapter 3 that, after the discovery that parity conservation is violated, it was conjectured that the combination of charge and parity (CP) reversal is really the conserved quantity. Our story left off there, implying that CP conservation is a good principle of physics. I intentionally neglected to say, however, that the full story had not yet been told. It is not that I was trying to be deceptive, but much needed to be learned in the ensuing chapters before the truth could be appreciated.

In postponing further discussion of CP violation, I was following a strategy often used in teaching the physics of simple motion. Newton and Galileo devised the first physical laws of motion three hundred years ago. Today, even though we know that Einstein's theory of relativity provides a more complete theory of the motion of objects, his theory is never taught first. It is a fact that most ordinary motion is completely described without invoking the complex theory of relativity. Even advanced space travel to the moon and between stars does not require the laws of relativity. The crew of Apollo 8, the first manned spacecraft to fly to the moon, remarked that it was Newton who was really flying the ship. Teachers have therefore found it best to begin with the simple theory of Newton and then, after about a year of introductory physics, to begin to describe relativity. Similarly, I oversimplified descriptions in earlier chapters, saving CP violation for now.

Before continuing the story of CP violation, I first need to tell the story of two puzzling meson particles, first discovered in the 1950s,

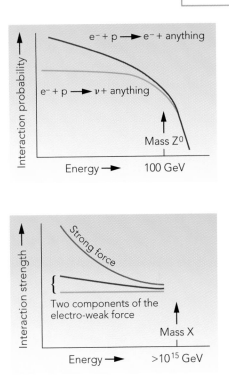

Two ways to view the unification of forces. Top: The unification of the electromagnetic and weak forces. Here charged current interactions (top curve) represent the weak force and neutral current interactions (bottom curve) represent the electromagnetic force (although imperfectly, since these interactions may be mediated by either of the two forces). The interaction probability for charged and neutral currents declines with increasing energy, based on observations made by the Zeus experiment at the HERA electron-proton collider. Above the mass of the Z^0 the two interactions are indistinguishable from each other. Bottom: The electromagnetic, strong, and weak nuclear forces decline in interaction strength as the energy of the interaction rises. A projection of these forces shows them merging at some very high energy and perhaps becoming one superforce.

composed of strange matter. The puzzle? These two particles have identical mass, but different parities. One meson decays into two pions and therefore has positive parity; the other meson decays into three pions and therefore demonstrates negative parity. Scientists at first agreed that these mesons were two different particles. Their assumption seemed reasonable since no other particle was known to have different parity states for its decay. But when physicists discovered that parity could be violated in weak decays, it was conjectured that these two mesons are, in fact, the same particle. Today this particle is called the kaon, K^+; it is composed of the up quark and antistrange quark, $u\bar{s}$, while its antiparticle, the K^-, is composed of the antiup quark and the strange quark, $\bar{u}s$.

We also know that there are two neutral kaons, K^0 and $\overline{K^0}$ particles, which mix between themselves, constantly converting back and forth between the two states as they propagate through space. Thus at different times an individual meson is either a $\bar{d}s$ or a $d\bar{s}$ combination of quarks. This mixing of neutral kaon particles is possible because of the special property of the down and strange quarks, discussed in Chapter 6, according to which the strange quark sometimes acts like the down quark and vice versa. However, what we actually see in nature are not the K^0 and $\overline{K^0}$ in pure form, but two types of particles, each of which represents a different probability of finding a K^0 or $\overline{K^0}$ when measured. These particles are said to be in either a K_1 state or a K_2 state. When a particle in the K_1 state is subjected to charge reversal and a parity inversion, it gives back the initial K_1 state. But when a particle in the K_2 state is subjected to the same charge and parity inversion, it returns a $-K_2$ state. What all this boils down to is that the existence of two different probabilities of finding a K^0 or a $\overline{K^0}$ gives rise to two different observable neutral kaons. Both neutral kaons are composed of two quarks, the strange quark and the down quark, and both vary back and forth in time between a $d\bar{s}$ and a $\bar{d}s$ mixture. The particles are distinguished from each other by each having a different parity, like the K^+, and also a different lifetime. These two kinds of neutral kaons are called the K-long and K-short particles, to distinguish between the long- and short-lived versions.

Another unusual feature of the neutral kaon particle is its capability for *regeneration*. Imagine that we've created a beam of neutral kaon particles and made it travel over a sufficiently long distance, for a sufficiently long time, so that all the short-lived kaons decayed away. We are thus left with a beam consisting only of K-long particles. Suppose that we aim the K-longs at a thin plate of matter, and then analyze the beam of particles just after its encounter with this material. We would find that the beam was again composed of long- and short-lived kaons. Through strong-force interactions with matter, some of the long-lived kaons convert back into the short-lived form (hence the term "regeneration"). Until recently the set of quarks composing the neutral kaon was the only system of quarks that was known to experience regeneration. Now, with the new higher-energy colliders, like Fermilab's Tevatron or CERN's LEP colliders, physicists have observed similar regeneration in the B^0 meson, although a sufficient number of events are not yet available for detailed studies. It was in the course of trying to study the kaon system's capacity for regeneration that physicists would find evidence for CP violation.

At Brookhaven National Laboratory in 1964, an experiment being run by James Christenson, Jim Cronin, Val Fitch, and Rene Turlay discovered something unusual. The long-lived kaon, which normally decays into three pions, was found to decay into two pions at the rate of one in a thousand. This simply would not happen if CP was conserved. Even the experimenters questioned their results and tried to find another explanation. They checked to see whether the material in their beam line were to blame, but were clearly able to identify a quantum mechanical interference effect that conclusively proved that the long- and short-lived kaons were not interacting with matter in the path of the beam. The only explanation left was that CP was not conserved.

More than thirty years have passed since the discovery of CP violation, and a satisfactory explanation of the phenomenon has yet to be found. Although CP violation occurs only rarely, making it hard to study, many experimenters continue to search for clues to its origin, while theorists continue to scrutinize the theory.

Shortly after the discovery of CP violation, the theorist Lincoln Wolfenstein came up with an intriguing and plausible explanation. He suggested that CP violation in the neutral kaon system is caused by a new force of nature. His theory, called *superweak*, gives a prediction for the ratio of neutral kaon decays, K-long and K-short, into their forbidden CP-violating mode. The standard model of particle physics makes its own predictions for this ratio. Both ratios can be checked by making comparisons with experimental results.

This investigation has proved frustrating, however, because the kaon system alone provides very few clues, and it is still the only known system for which we are certain that CP violation occurs. If the standard model is correct, then other systems of quarks, such as the B^0 meson made of $d\bar{b}$ and the $\overline{B^0}$ antimeson composed of $\bar{d}b$, should exhibit the same properties as the kaon; but thus far too few B^0 particles have been produced to permit a search for CP violation. Wolfenstein's superweak theory, on the other hand, does not necessarily expect other systems of quarks to have a similar CP-violating effect. By studying systems similar to the K^0, it may be possible to learn something more conclusive about the mystery of CP violation. At present there is great excitement about the building of a B^0 factory at SLAC in California that will produce thousands of B^0 particles a day, enough to search for CP violation in this system.

The new KTeV experiment at Fermilab is searching for the origin of CP violation. Its large, high-precision magnet and its precision tracking and particle-identification detectors are visible on the level below. In addition, the experiment has a high-rate electromagnetic calorimeter with spectacular energy resolution. All these instruments are working together to measure neutral kaon decays.

With the downfall of the CP symmetry of nature, we are left with only CPT symmetry: the combination of charge reversal, parity (spatial inversion), and time reversal—heralded as the ultimate symmetry of nature. Many experiments have looked for evidence that this symmetry is violated, but none have found it. Nevertheless, the search continues. Another recent theory evokes Higgs field coupling to explain the origin of CP violation, something that could actually be studied when the Higgs particle is produced at a collider.

Matter versus Antimatter

CP violation may have an important astronomical implication. Astronomers have found no evidence that matter is in contact with antimatter anywhere in the universe. Although we see no signs of antimatter, our universe might still be composed equally of matter and antimatter if the two forms of matter had been separated into different parts of the universe in the very distant past. But if that were the case, then somewhere in the universe there should be a boundary where matter is in contact with antimatter. Astronomers have searched unsuccessfully for the 1 MeV peak in the X-ray spectrum, or the 2 GeV peak in the gamma-ray spectrum, that would be generated at the boundary by electron-positron or proton-antiproton annihilation. Their data still leaves us with the question: Why is our universe made only of matter?

The very early universe, before the strong force decoupled from electro-weak interactions at 10^{-36} second, was a quark-gluon soup. During this stage, the strongly interacting quarks and antiquarks would have annihilated. However, in this brief 10^{-36} second, CP violation processes are believed to have caused a small imbalance to occur, and the universe came to have a slight excess of matter over antimatter. All of the matter remaining today is supposedly here because this slight violation of CP symmetry permitted some processes of antimatter conversion into matter while restricting the inverse reactions. Although this dynamic occurs in the kaon system, the CP violation generating an excess of matter over antimatter must have been very different because the meson had not formed in the universe by that time. The CP violation in the early universe could have even been of a completely different nature from the CP violation observed in modern-day laboratory experiments because it would also involve the grand unified theory propagator particles X and Y. Another explanation of our matter-dominated universe could be that the antimatter did not annihilate after all, but went on to form antiprotons. Some physicists think antiprotons may be unstable and decay away, just like the free neutron. Currently all we can say about the lifetime of the antiproton is that it must be greater than a million years.

There is one large problem with the CP violation theory of matter over antimatter. If CP violation is really the origin of matter in the universe, then at some time in the very distant past the rate of CP violation must have been tens or hundreds of times higher than its currently measured value. There is simply too much matter in our universe to be attributed to CP violation at the strengths we have been able to measure. But, as we saw in Chapter 9, the very early universe was *not* dominated by normal matter or even strange matter; rather, it contained a lot of third-generation quarks and leptons. The rate problem might be solved if CP violation could be shown to be more copious in third-generation quark particles than in second-generation quark particles. The B^0 system may offer us our best chance to explore CP violation in the third-generation quark. Here, then, is another reason for the great interest in building the new collider and experiments at SLAC. Unfortunately, as already noted, the early universe at the time of the production of matter over antimatter did not even have heavy mesons, but was a soup of freely interacting quarks and gluons such as the heavy ion colliders are trying to produce.

Recent speculation is that CP violation could also explain the presence of matter in our universe if matter at high nuclear densities exhibits higher rates of CP violation. A good place to test this hypothesis is in heavy ion collisions, which can now produce nuclear matter of up to six times normal density. If, in these interactions, a higher rate of CP violation is found, then it would prove that the very early universe did indeed have a much higher rate of CP violation. An alternative astronomical explanation is that it might have taken the very early universe much longer to expand than current theories predict, providing more time for the CP violation to generate an excess of matter. In this case, however, our thermal history of the universe would have to be rewritten. Furthermore, astronomers are more certain of the current estimates of the expansion time than of anything else they can estimate about the early universe. A way to experimentally confirm the expansion time would be to measure the thermal energy of the primordial neutrinos left behind by the early universe, but such a measurement is beyond our current technical abilities.

. . .

Talking about the possible new discoveries to be made in the future has been an interesting way to end our discussion of particle physics. Of the discipline's many unanswered questions, one of the most eagerly discussed concerns the Higgs field: Does the field really exist and can a theory be devised to permit us to understand how particles came to have such varying masses? Or will the particles we believe to be elementary simply be shown to be composed of even more fundamental particles?

Only experiments, guided by theory, of course, will be able to answer these questions.

Other unanswered questions concern symmetry. After parity was found to be violated through the weak force, physicists presumed that CP, the combination of charge and parity symmetry, was the remaining valid symmetry of nature; but in the mysterious neutral kaon system this, too, was found to be violated. The same laws of elementary particle physics that cause this phenomenon, although hardly understood, may have created the imbalance of matter over antimatter, again revealing the interconnection between the small elementary particles and the large-scale structure observed in the universe. Now physicists are attempting to determine the universal validity of CPT symmetry, the combination of charge, parity, and temporal symmetry, as they continue their endeavor to learn more about all the subatomic particles, including the neutrino.

Epilogue

The neutrino remains elusive—still difficult to detect, still of unknown mass. Although physicists have long known how to measure the neutrino's mass, technical limitations have prevented these measurements from achieving a sufficient level of accuracy. Nor can we predict the mass of the neutrino, or of any other particle, from theory alone. The hope that the Higgs theory may one day permit us to calculate particle masses from simple principles is the reason for all the excitement that this theory has generated.

A recent controversy illustrates some of the pitfalls of mass measurement. During 1992 excitement spread through the physics community at the report of a new type of neutrino, observed in ordinary beta decay, with a mass of 17 keV/c^2. Many prominent physicists were claiming to have confirmed this observation, but just as many others were skeptical. The critics could not accept that such an easy observation had not been made much earlier with the equipment available back in the 1950s, '40s, or even '30s.

The new experiment used a type of detector made out of silicon. As it turned out, when the new observation was analyzed with more conventional gas-filled multiwire detectors, which were also expected to be sensitive to such a heavy neutrino, the signal for the neutrino was not present. In the end physicists realized that the 17 keV/c^2 neutrino was an artifact of the silicon detector. This confusing story took two and a half years to unfold, as physicists using the new type of detector repeatedly confirmed the mistaken result. Even though all the checks and balances of the scientific process were in place, an incorrect result continued to be accepted simply because the experimenters were unfamiliar with the subtleties of a new detector. Experiments now in the planning stage will permit the neutrino mass to be measured down to values as low as 4 eV/c^2. However, like the best direct mass measurements now available, which can predict only that the mass is less than 9 eV/c^2, they will not tell us whether the neutrino is massless.

Its elusiveness is but one of the qualities of the neutrino that continue to tantalize scientists. The neutrino is both a puzzle and the possible solution to a puzzle. It could be the answer to the chief conundrum now bedeviling cosmologists: What constitutes the missing dark matter of the universe? We ended Chapter 2 by concluding that the electron-type neutrino was not a good candidate for the missing matter: its mass is too small to determine our cosmological fate. But the neutrino comes in more than one variety: there is the muon neutrino, also too small in mass, and the as yet undetected tau neutrino mentioned briefly in Chapter 9. Today some particle theorists suspect that this third type of neutrino may be the key to the missing matter in the universe. The tau neutrino was produced copiously in the first microsecond of the big bang, and its mass, though poorly measured, seems to be greater than the masses of the other two neutrino types combined. This particle just may be the dark matter that could control the universe's fate.

Scientists now believe that as the universe ages matter will continue to clump together into supermassive objects such as galaxies, quasars, and black holes. If there is enough matter to make the universe collapse, then all of space will contract into a singularity, and the universe may be born anew. If there is not enough matter, then the universe will continue to expand and cool, the stars will die and matter decay. According to one scenario, the universe will be left at the end with only three constituents: black holes, the photons making up the background radiation, and neutrinos. If Stephen Hawking's theory that black holes evaporate is correct, the surviving constituents will instead be electrons and positrons, produced as black holes evaporate, as well as neutrinos and the background radiation of photons. The neutrino endures in both scenarios. In its apparent immortality, the neutrino proves to be special once again.

Perhaps the most perplexing characteristic of the neutrino is the one that sets it apart from every other particle—its persistence in a state of left-handed spin despite the fact that every other particle has both left- and right-handed spin. Why is the spin state of the neutrino so different from that of every other known particle? Does the reason lie in its perhaps zero mass? Or in the fact that of all the particles of matter shown in the table on page 97 it is the only one that has no charge? Or are its unique qualities the result of some still unknown aspect of the force through which the neutrino is created?

Eventually scientists will learn more about the neutrino's mysterious aspects, either through their investigations of the neutrino itself or through their explorations of the many other questions that excite our curiosity in the world of particle physics. At the same time investigations of the neutrino will continue to provide valuable information about the structure of matter in the universe we live in. Observations of neutrino interactions with quarks might yield the data that physicists need to incorporate the strong force with that of electro-weak interactions. And if the current attempts to determine the origin of CP violation in the kaon or B^0 system fail, then the next best hope might just lie in the long-baseline neutrino oscillation experiments. The neutrino will assuredly remain a valuable tool for studying particle interactions as well as an interesting subject of study in its own right.

Further Readings

Biagioli, Mario. *Galileo Courtier,* University of Chicago Press, Chicago, 1993.

An excellent account of how modern science was conducted in its earliest days. It will interest those curious about the birth of modern science and the significant role that Galileo played. Perhaps its most amazing insight is to show how little the politics of the academic world have changed. Biagioli covers in depth the historical material touched upon in Chapters 2 and 8 of *The Elusive Neutrino.*

Feynman, Richard. *The Character of a Physical Law,* MIT Press, Cambridge, Massachusetts, 1967.

An account of how scientific laws are formed. Feynman gives the reader a sense of what is important to scientific methodology, although he gives too much credit to scientists as brilliant naturalists. His book is a good introduction to quantum mechanics and its subtle mysteries. It complements the scientific lessons of this book, and it provides a nice background to the quest for further scientific knowledge presented in Chapter 10.

Lederman, Leon M., and David N. Schramm. *From Quarks to the Cosmos: Tools of Discovery,* W. H. Freeman, New York, 1995.

An excellent book that describes the workings of many of the tools scientists use in both particle physics and astronomy experiments. *From Quarks to the Cosmos* covers more thoroughly the early days of particle physics presented in Chapters 1, 3, and 5. It is interesting to read Schramm himself explain the importance of particle physics experiments to cosmology, present in Chapter 9 of *The Elusive Neutrino.*

Pauli, Wolfgang. *Writings on Physics and Philosophy,* C. P. Enz and K. von Meyenn (eds.), Springer-Verlag, Heidelberg, 1994.

The only opportunity left for us to learn directly from Pauli what he thought about the neutrino, particle physics, and quantum mechanics. A valuable book for anyone who wishes a scholarly understanding of the debates of the early days of elementary particle physics. Pauli's writings provide additional insight into the material presented in Chapters 1, 3, 4, and 6 of *The Elusive Neutrino.*

Polkinghorne, J. C. *The Particle Play,* W. H. Freeman, San Francisco, 1979.

A book that opened up the world of particle physics to me when I read it in high school. Polkinghorne shows how the physics before 1930, although deceptively simple, presented challenges as difficult as those of today. *The Particle Play* complements the material presented in Chapters 5, 6, 7, and 10.

Wald, Robert M. *Space, Time, and Gravity: The Theory of the Big Bang and Black Holes,* 2d ed., University of Chicago Press, Chicago, 1992.

A wonderful introduction to the adventure of scientific thought. Wald's book complements the material presented in Chapters 2, 8, and 9.

Weinberg, Steven. *The First Three Minutes,* 2d ed., Basic Books, New York, 1993.

An excellent account of the big bang, including both the particle physics and the cosmology of the process. It is aimed at the general public despite the massive quantities of numbers, which, if ignored, leaves one with a book that is both entertaining and educational. Weinberg's book is an extended discussion of the early universe, a subject which is found in Chapter 9 of this book.

Sources of Illustrations

CHAPTER 1

Facing page 1: R. Williams, STSCI, NASA
Page 3: Fermilab Visual Media Services
Page 4: CERN Photo
Page 5: The Corning Museum of Glass
Page 6: Archives Curie and Joliot-Curie, Paris
Page 9: Science Source/Photo Researchers
Page 14: The Pauli Archives, Geneva
Page 16: English translation of Fermi's letter from E. Amaldi, "From the Discovery of the Neutron to the Discovery of Nuclear Fission," *Physics Reports,* vol. III, September 1984. Elsevier Science Publishers, Amsterdam
Page 18: Elsevier Science Publishers, Amsterdam, from E. Amaldi, "From the Discovery of the Neutron to the Discovery of Nuclear Fission," in *Physics Reports,* vol. III, September 1984.

CHAPTER 2

Page 20: Ian Gatley, Michael Merrill/NOAO, Roger Ressmeyer/© Corbis
Page 22: © Anglo Australian Observatory 1980, photograph by D. Malin
Page 25: Spectrum by jtalbot@achilles.net © Digitized star spectra courtesy of the Astronomical Data Center and the National Space Science Data Center through the World Data Center A for Rockets and Satellites
Page 31: © Anglo-Australian Observatory, photo by D. Malin
Page 32: The three theoretical curves are from Fermi's original 1934 paper: E. Fermi, *Z. Physik,* vol. 88 (1934), page 161. The data on the right are from the University of Zurich result published in: M. Fritschi et al., *Physics Letters B,* vol. 173, page 485

CHAPTER 3

Page 36: Lawrence Berkeley National Laboratory
Page 38: Yukawa Institute for Fundamental Physics, Kyoto, Japan
Page 47: Columbia University, Department of Physics
Page 48: Adapted from a figure in: C.-S. Wu et al., *Physical Review,* vol. 105 (1957), page 1413

Page 52: Based on a description and simpler sketch in the original paper: M. Goldhaber et al., *Physical Review,* vol. 109 (1958), page 1015

CHAPTER 4

Page 54: Leon Golub, SAO and IBM Research
Page 57: Rutherford's letter was reprinted in: E. Amaldi, "From the Discovery of the Neutron to the Discovery of Nuclear Fission," *Physics Reports,* vol. III, September 1984. Elsevier Science Publishers, Amsterdam
Pages 60–61: Einstein's letter is in the President Roosevelt Archives, Hyde Park, New York
Page 62: Argonne National Laboratory photo, sketch by Melvin A. Miller
Page 66: Sovfoto/Eastfoto Agency
Page 67: NASA, Corbis & Roger Ressmeyer/© Corbis
Page 70: Based on the theoretical work of: J. N. Bahcall and R. K. Ulrich, *Reviews of Modern Physics,* vol. 60 (1988), page 297
Page 71: Brookhaven National Laboratory
Page 72: Courtesy of Raymond Davis
Page 74: Max Planck Institut für Kernphysik

CHAPTER 5

Page 78: CERN Photo
Page 82: Collection of E. Segrè. Courtesy of C. F. Powell
Page 85: Based on a figure in: Leon M. Lederman, "The Two-Neutrino Experiment," *Scientific American,* March 1963, pages 66–67. © 1963 by Scientific American, Inc. All rights reserved
Page 87 (top and bottom): Brookhaven National Laboratory
Page 96: Based on a similar drawing in: M. Aguilar-Benitez et al., "Review of Particle Properties," *Physics Letters B,* vol. 239, April 1990

CHAPTER 6

Page 100: CERN Photo
Page 103: Dr. Thomas Ypsilantis
Page 104: Adapted from two figures in: Emilio Segrè and

Clyde E. Wiegand, "The Antiproton," *Scientific American*, vol. 194 (no. 6), June 1956. © 1956 by Scientific American, Inc. All rights reserved

Page 106: Dr. Thomas Ypsilantis

Page 107: Lawrence Berkeley National Laboratory

Page 108: The Ettore Majorana Center, Erice, Italy

Page 110 (bottom): Based on a graph by S. R. Elliott, A. Hahn, and M. K. Moe, in *Physical Review Letters*, vol. 59 (1987), page 1649

Page 112: H. V. Klapdor-Kleingrothaus, Max Planck Institut für Kernphysik

Page 115: CERN Photo

Page 116: Fred M. Rick, Los Alamos National Laboratory

Page 117: Courtesy of David S. Ayres, Argonne National Laboratory

CHAPTER 7

Page 120: CERN Photo

Page 122: CERN Photo

Page 123: CERN Photo

Page 125: CERN Photo

Page 126: CERN Photo/Courtesy of Coline Musset

Page 128: Argonne National Laboratory photo

Page 137: From David B. Cline, Carlo Rubbia, and Simon van der Meer, "The Search for Intermediate Vector Bosons," *Scientific American*, March 1982. © 1982 by Scientific American Inc. All rights reserved

Page 138: CERN Photo

Page 139: CERN Photo

Page 140: CERN Photo

CHAPTER 8

Page 142: Dr. Christopher Burrows, ESA/STSCI and NASA

Page 145: Courtesy of H. Tananbaum, Harvard-Smithsonian Center for Astrophysics

Page 146: S. L. Snowden et al., 1997, *Astrophysical Journal*, in press, and ROSAT, PSPC, MMPE

Page 147 (left and right): Carl Fichtel, Goddard Space Flight Center, NASA

Page 149: Courtesy of Carl Fichtel, Goddard Space Flight Center, NASA

Page 150: NASA, STSCI

Page 152: Yerkes Observatory, photo by Stephan M. Kent, University of Chicago

Page 153: Joe Stancampiano © National Geographic Society Image Collection

Page 154: Daniel Sinclair IMB collaboration

Page 155: Based on data published in: R. M. Bionta et al., *Physical Review Letters*, vol. 58 (1987), page 1494; and K. Hirata et al., *Physical Review Letters*, vol. 58, page 490

Page 158: Courtesy of David S. Ayres, Argonne National Laboratory

CHAPTER 9

Page 160: CERN Photo

Page 168: Dr. Gary Steigman, Professor of Physics and Astronomy, Ohio State University

Page 169: CERN Photo

Page 170: Based on a drawing provided by CERN

Page 171: CERN Photo

Page 172: CERN Photo

Page 173: Data points (from L3) published in: B. Adeva et al., *Physics Letters B*, vol. 249 (1990), page 341

Page 175: Fermilab Visual Media Services

Page 177: Fermilab Visual Media Services

Page 178: Fermilab Visual Media Services

CHAPTER 10

Page 180: CERN Photo

Page 183: Photo by Robert Palmer

Page 185: CERN Photo

Page 189: CERN Photo

Page 195: Dr. Nickolas Solomey

Many of the line drawings are based on original black-and-white renderings by Elizabeth Pod; the drawings on pages 59 and 158 are based on renderings by Richard Armstrong. Hudson River Graphics produced the final versions that appear in the book.

Index

Selected Books in the Scientific American Library Series

If you would like to purchase additional volumes in the Scientific American Library, please send your order to:

Scientific American Library
41 Madison Avenue
New York, NY 10010